Enlightening Symbols

Also by Joseph Mazur

*Euclid in the Rainforest: Discovering Universal Truth
in Logic and Math* (2006)

*Zeno's Paradox: Unraveling the Ancient Mystery
behind the Science of Space and Time* (2007)

*What's Luck Got to Do with It? The History, Mathematics,
and Psychology of the Gambler's Illusion* (2010)

Edited

Number: The Language of Science (2007)

Enlightening Symbols

A Short History of Mathematical Notation
and Its Hidden Powers

Joseph Mazur

Princeton University Press
Princeton and Oxford

Published by Princeton University Press, 41 William Street,

Princeton, New Jersey 08540

In the United Kingdom: Princeton University Press, 6 Oxford Street,

Woodstock, Oxfordshire OX20 1TW

press.princeton.edu

Library of Congress Cataloging-in-Publication Data

Mazur, Joseph.

Enlightening symbols : a short history of mathematical notation and its hidden

powers / Joseph Mazur.

pages cm

Includes bibliographical references and index.

ISBN 978-0-691-15463-3 (hardcover : alk. paper) 1. Mathematical

notation—History. I. Title.

QA41.M39 2014

510.1′48—dc23

2013028571

British Library Cataloging-in-Publication Data is available

This book has been composed in Minion and Candida

Printed on acid-free paper. ∞

Printed in the United States of America

3   5   7   9   10   8   6   4   2

To my big brother, Barry,
who taught me from *0*

Contents

Introduction

A mathematician, a musician, and a psychologist walked into a bar ...

Several years ago, before I had any thoughts of writing a book on the history of symbols, I had a conversation with a few colleagues at the Cava Turacciolo, a little wine bar in the village of Bellagio on Lake Como. The psychologist declared that symbols had been around long before humans had a verbal language, and that they are at the roots of the most basic and primitive thoughts. The musician pointed out that modern musical notation is mostly attributed to one Benedictine monk Guido d'Arezzo, who lived at the turn of the first millennium, but that a more primitive form of symbol notation goes almost as far back as Phoenician writing. I, the mathematician, astonished my friends by revealing that, other than numerals, mathematical symbols—even algebraic equations—are relatively recent creations, and that almost all mathematical expressions were rhetorical before the end of the fifteenth century.

"What?!" the psychologist snapped. "What about multiplication? You mean to tell us that there was no symbol for 'times'?"

"Not before the sixteenth... maybe even seventeenth century."

"And equality? What about 'equals'? the musician asked.

"Not before... oh... the sixteenth century."

"But surely Euclid must have had a symbol for addition," said the psychologist. "What about the Pythagorean theorem, that thing about adding the squares of the sides of a right triangle?"

"Nope, ...no symbol for 'plus' before the twelfth century!"

A contemplative silence followed as we sniffed and sipped expensive Barolo.

As it turned out, I was not correct. And far, far back in the eighteenth century BC, the Egyptians had their hieroglyphical indications of addition and subtraction in glyphs of men running toward or away from amounts to be respectively added or subtracted. And from time to time, writers of mathematical texts had ventured into symbolic expression. So there are instances when they experimented with graphic marks to represent words or even whole phrases. The *Bakhshâlî* manuscript of the second century BC records negative numbers indicated by a symbol that looks like our plus sign. In the third century, Diophantus of Alexandria used a Greek letter to designate the unknown and an arrow-like figure pointing upward to indicate subtraction. In the seventh century, the Indian mathematician Brahmagupta used a small black dot to introduce the new number we now call "zero." And symbols were timidly beginning to find their way into mathematics by the second half of the fifteenth century. Of course, for ages, there have been the symbols that we use to designate whole positive numbers.

That night at the *enoteca*, I didn't know that my estimate for the adoption of symbols was premature by several centuries. Sure, Diophantus in the third century had his few designations; however, before the twelfth century, symbols were not used for operational manipulation at the symbolic level—not, that is, for purely symbolic operations on equations. Perhaps I should have pushed the edge of astonishment to claim, correctly, that *most* mathematical expressions were rhetorical before the sixteenth century.

Ever since that conversation, I have found that most people are amazed to learn that mathematics notation did not become really symbolic before the sixteenth century. We must also wonder: What was gained by algebra taking on a symbolic form? What was lost?

Traced to their roots, symbols are a means of perceiving, recognizing, and creating meaning out of patterns and configurations drawn from material appearance or communication.

The word "symbol" comes from the Greek word for "token," or "token of identity," which is a combination of two word-roots, *sum* ("together") and the verb *ballo* ("to throw"). A more relaxed interpretation would be "to put together." Its etymology comes from an ancient way of proving one's identity or one's relationship to another. A stick or bone would be broken in two, and each person in the relationship would be given one piece. To verify the relationship, the pieces would have to fit together perfectly.

On a deeper level, the word "symbol" suggests that, when the familiar is thrown together with the unfamiliar, something new is created. Or, to put it another way, when an unconscious idea fits a conscious one, a new meaning emerges. The symbol is exactly that: meaning derived from connections of conscious and unconscious thoughts.

Can mathematical symbols do that? Are they meant to do that? Perhaps there should be a distinction between symbols and notation. Notations come from shorthand, abbreviations of terms. If symbols are notations that provide us with subconscious thoughts, consider "+." Alone, it is a notation, born simply from the shorthand for the Latin word *et*. Yes, it comes from the "t" in *et*. We find it in 1489 when Johannes Widmann wrote *Behende und hubsche Rechenung auff allen Kauffmanschafft* (*Nimble and neat calculation in all trades*). It was meant to denote a mathematical operation as well as the word "and."

Used in an arithmetic statement such as $2 + 3 = 5$, the "+" merely tells us that 2 and 3 more make 5. But in the context of an algebraic statement such as $x^2 + 2xy + y^2$ it generally means more than just "x^2 and $2xy$ and y^2." The mathematician sees the +'s as the glue to form the perfect square $(x + y)^2$. Now surely the same mathematician would just as well see the "and" as the glue. Perhaps it may take a few more seconds to recognize the perfect square, but familiar symbols habitually provide useful associations when we are looking at one object while knowing that it has another useful form.

A purist approach would be to distinguish symbolic representation from simple notation. I have a more generous slant; numerals and all nonliteral operational notation are different, but still considered symbols, for they represent things that they do not resemble.

Read the statement $2 + 3 = 5$ again. It is a complete sentence in mathematics, with nouns, a conjunction, and a verb. It took you about a second to read it and continue on. Unaware of your fact-checking processes, you believe it for many reasons, starting from what you were told as a young child and ending with a mountain of corroborating evidence from years of experience. You didn't have to consciously search through your mental library of truthful facts to *know* that it is true.

Yet there is a distinct difference between the writer's art and the mathematician's. Whereas the writer is at liberty to use symbols in ways that contradict experience in order to jolt emotions or to create states of mind with deep-rooted meanings from a personal life's journey, the mathematician cannot compose contradictions, aside from the standard argument that establishes a proof by contradiction. Mathematical symbols have a definite initial purpose: to tidily package complex information in order to facilitate understanding.

Writers have more freedom than mathematicians. Literary symbols may be under the shackles of myth and culture, but they are used in many ways. Emily Dickinson never uses the word "snake" in her poem "A Narrow Fellow in the Grass," thereby avoiding direct connections with evil, sneakiness, and danger, though hinting all the same. Joseph Conrad invokes all the connotations of slithering, sneaky evil in *Heart of Darkness* when describing the Congo River as "an immense snake uncoiled, with its head in the sea." It is also possible that a writer may use the word "snake" innocently, in no way meaning it as something unsuspected, crafty, or dangerous. It could be simply a descriptive expression, as in "the river wound around its banks like a snake." The writer may intend to invoke an image in isolation from its cultural baggage. This is tough—perhaps impossible—to do with words or expressions that are so often used figuratively.

Mathematicians use a lemma (a minor theorem used as a stepping stone to prove a major theorem) called the "snake lemma," which involves a figure called the "snake diagram"—it doesn't mean that there is anything sinister, crafty, or dangerous within, but rather that the figure just happens to look like a snake, again just a graphic description.

Human-made symbols of mathematics are distinct from the culturally flexible, emotional symbols found in music or from the metaphorical symbols found in poems. However, some also tend to evoke subliminal, sharply focused perceptions and connections. They might also transfer metaphorical thoughts capable of conveying meaning through similarity, analogy, and resemblance, and hence are as capable of such transferences as words on a page.

In reading an algebraic expression, the experienced mathematical mind leaps through an immense number of connections in relatively short neurotransmitter lag times.

Take the example of π, the symbol that every schoolchild has heard of. As a symbol, it is a sensory expression of thought that awakens intimations through associations. By definition, it means a specific ratio, the circumference of a circle divided by its diameter. As a number, it is approximately equal to 3.14159. It masquerades in many forms. For example, it appears as the infinite series

$$\pi = \frac{4}{1} - \frac{4}{3} + \frac{4}{5} - \frac{4}{7} + \frac{4}{9} \cdots,$$

or the infinite product

$$\pi = 2 \cdot \frac{2}{1} \cdot \frac{2}{3} \cdot \frac{4}{3} \cdot \frac{4}{5} \cdot \frac{6}{5} \cdot \frac{6}{7} \cdot \frac{8}{7} \cdot \frac{8}{9} \cdots,$$

or the infinite fraction

$$\pi = \cfrac{4}{1 + \cfrac{1^2}{3 + \cfrac{2^2}{5 + \cfrac{3^2}{7 + \ldots}}}}.$$

It frequently appears in both analytical and number theoretic computations. When she sees π in an equation, the savvy reader automatically knows that something circular is lurking behind. So the symbol (a relatively modern one, of course) does not fool the mathematician who is familiar with its many disguises that unintentionally drag along in the mind to play into imagination long after the symbol was read.

Here is another disguise of π: Consider a river flowing in uniformly erodible sand under the influence of a gentle slope. Theory predicts that over time the river's actual length divided by the straight-line distance between its beginning and end will tend toward π. If you guessed that the circle might be a cause, you would be right.

The physicist Eugene Wigner gives an apt story in his celebrated essay, "The Unreasonable Effectiveness of Mathematics in the Natural Sciences."[1] A statistician tries to explain the meaning of the symbols in a reprint about population trends that used the Gaussian distribution. "And what is this symbol here?" the friend asked.

"Oh," said the statistician. "This is pi."

"What is that?"

"The ratio of the circumference of the circle to its diameter."

"Well, now, surely the population has nothing to do with the circumference of the circle."

Wigner's point in telling this story is to show us that mathematical concepts turn up in surprisingly unexpected circumstances such as river lengths and population trends. Of course, he was more concerned with understanding the reasons for the unexpected connections between mathematics and the physical world, but his story also points to the question of why such concepts turn up in unexpected ways within pure mathematics itself.[2]

The symbol π had no meaning in Euclid's *Elements* (other than its being the sixteenth letter of the ancient Greek alphabet), even though the *Elements* contained the proof of the hard-to-prove fact that the areas of any two circles are to one another as the squares on their diameter.[3] The exceptionality of Greek mathematical thinking is in conceiving that there are universal truths that could be proven: that any circle is bisected by any of its diameters, that the sum of angles in any triangle is always the same constant number, that only five regular solids can exist in three dimensions. In book II, proposition 4, Euclid showed us how to prove what we might today think of as simple algebraic identities, such as $(a + b)^2 = a^2 + b^2 + 2ab$, but you will not find any algebraic symbols indicating powers (those little raised numbers that tell how many times to multiply a number by itself) or addition in his proposition or proof because his statements and proofs were, on the one hand, geometrical and, on the other, entirely in narrative form.

Diophantus of Alexandria was born more than five hundred years after Euclid. His great work, *Arithmetica*, gave us something closer to algebraic solutions of special linear equations in two unknowns, such as $x + y = 100$, $x - y = 40$. He did this not by using the full power of symbols, but rather by syncopated notation—that is, by the relatively common practice of the time: omitting letters from the middle of words. So his work never fully escaped from verbal exposition.[4] It was the first step away from expressing mathematics in ordinary language.

It is possible to do all of mathematics without symbols. In general, articles of law contain no symbols other than legalese such as "appurtenances," "aforesaid," "behoove"—words that few people would dream of using in anything other than a legal document. By tradition, and surely by design, law has not taken the symbolic road to precision. Words in a natural language such as English or Latin can present tight meaning, but almost never ironclad precision the way symbolic algebra can. Instead, written law relies heavily on intent, and expects loopholes to be found by those clever people who use them.

Imagine what mathematics would be like if it were still entirely rhetorical, without its abundance of cleverly designed symbols. Take a passage in al-Khwārizmī's *Algebra* (ca. 820 AD) where even the numbers in the text are expressed as words:

> If a person puts such a question to you as: "I have divided ten into two parts, and multiplying one of these by the other the result was twenty-one;" then you know that one of the parts is thing, and the other is ten minus thing.[5]

We would write this simply as: $x(10 - x) = 21$.

The language of the solution, as al-Khwārizmī wrote it, was specific to the question. There may have been a routine process, some algorithm, lurking behind the phrasing, but it would have taken work to bring it out, since al-Khwārizmī's *Algebra* is not particularly representative of the mathematics of his period.

Privately things may have been different. Thinking and scratch work probably would have gone through drafts, just as they do today. I have no way of knowing for sure, but I suspect that the solution was first probed on some sort of a dust board using some sort of personal notation, and afterward composed rhetorically for text presentation.

The sixth-century prolific Indian mathematician-astronomer Aryabhatta used letters to represent unknowns. And the seventh-century Indian mathematician-astronomer Brahmagupta—who, incidentally, was the first writer to use zero as a number—used abbreviations for squares and square roots and for each of several unknowns occurring in special problems. Both Aryabhatta and Brahmagupta wrote in verse, and so whatever symbolism they used had to fit the meter. On seeing a dot, the reader would have to read the word for dot. This put limitations of the use of symbols.[6] A negative number was distinguished by a dot, and fractions were written just as we do, only without the bar between numerator and denominator.

Even as late as the early sixteenth century, mathematics writing in Europe was still essentially rhetorical, although for some countries certain frequently used words had been abbreviated for centuries. The abbreviations became abbreviated, and by the next century, through the writings of François Viète, Robert Recorde, Simon Stevin, and eventually Descartes, those abbreviations became so compacted that all the once-apparent connections to their origins became lost forever.

In mathematics, the symbolic form of a rhetorical statement is more than just convenient shorthand. First, it is not specific to any particular language; almost all languages of the world use the same notation, though possibly in different scriptory forms. Second, and perhaps most importantly, it helps the mind to transcend the ambiguities and misinterpretations dragged along by written words in natural language. It permits the mind to lift particular statements to their general form. For example, the rhetorical expression *subtract twice an unknown from the square of the unknown and add one* may be written as $x^2 - 2x + 1$. The symbolic expression might suggest a more collective notion of the expression, as we are perhaps mentally drawn from the individuality of $x^2 - 2x + 1$ to the general quadratic form $ax^2 + bx + c$. We conceive of $x^2 - 2x + 1$ merely as a representative of a species.

By Descartes's time at the turn of the seventeenth century, rhetorical statements such as

> The square of the sum of an unknown quantity and a number equals the sum of the squares of the unknown and the number, augmented by twice the product of the unknown and the number.

were written almost completely in modern symbolic form, with the symbol ∞ standing for equality:

$$(x + a)^2 \infty x^2 + a^2 + 2ax$$

The symbol had finally arrived to liberate algebra from the informality of the word.

As with almost all advances, something was lost. We convey modern mathematics mostly through symbolic packages, briefcases (sometimes suitcases) of information marked by symbols. And often those briefcases are like Russian *matrioshka* dolls, collections of nested briefcases, each depending on the symbols of the next smaller *matrioshka*.

There is that old joke about joke tellers: A guy walks into a bar and hears some old-timers sitting around telling jokes. One of them calls out, "Fifty-seven!" and the others roar with laughter. Another yells, "Eighty-two!" and again, they all laugh.

So the guy asks the bartender, "What's going on?"

The bartender answers, "Oh, they've been hanging around here together telling jokes for so long that they catalogued all their jokes by number. All they need to do to tell a joke is to call out the number. It saves time."

The new fellow says, "That's clever! I'll try that."

So the guy turns to the old-timers and yells, "Twenty-two!"

Everybody just looks at him. Nobody laughs.

Embarrassed, he sits down, and asks the bartender, "Why didn't anyone laugh?" The bartender says, "Well, son, you just didn't tell it right..."

Mathematicians often communicate in sequentially symbolic messages, a code, unintelligible to the uninitiated who have no keys to unlock those briefcases full of meaning. They lose the public in a mire of marks, signs, and symbols that are harder to learn than any natural language humans have ever created.

More often, in speaking, for the sake of comprehension, they relax their airtight arguments at the expense of mildly slackening absolute proof. They rely on what one may call a "generosity of verbal semantics," an understanding of each other through a shared essence of professional expertise and experience independent of culture.

However, even with a generosity of verbal semantics, something beyond absolute proof is lost. Mathematics, even applied mathematics, physics, and chemistry, can be done without reference to any physically imaginable object other than a graphic symbol. So the difference between the physicist's rhetorical exposition and the mathematician's is one of conceptualization.

That might be why physicists have an easier time communicating with the general public; they are able to give us accounts of "stuff" in this world. Their stuff may be galaxies, billiard balls, atoms, elementary particles of matter and strings, but even those imperceptible strings that are smaller than 10^{-35} of a meter in ten-dimensional space are imagined as stuff. Even electric and magnetic fields can be imagined as stuff. When the physicist writes a book for a general audience, she starts with the advantage of knowing that every one of her readers will have experience with some of the objects in her language, for even her most infinitesimal objects are imaginable "things."

A mathematician's elements are somewhat more intangible. The symbol that represents a specific number N is more than just a notational convenience to refer to an indeterminate number. These days it represents an object in the mind with little cognate reference to the world—in other words, the N is a "being" in the mind without a definite "being" in the world. So the nonphysical object has an ontology through cognition. The mind processes a modern understanding of a number—say, three—just as it does for any abstraction, by climbing a few rungs of increasing generality, starting at a definite number of things within the human experience: three sheep in the field, three sheep, three living things, three things, ... all the way up to "three-ness." Imagining physical objects decreases as generality increases. The mathematical symbol, therefore, is a visual anchor that helps the mind through the process of grasping the general through the particular.

———————+———————

This book traces the origins and evolution of established symbols in mathematics, starting with the counting numbers and ending with the primary operators of modern mathematics. It is chiefly a history of mathematical symbols; however, it is also an exploration of how symbols affect mathematical thought, and of how they invoke a wide range of enduring subconscious inspirations.

It is arranged in three parts to separate the development of numerals from the development of algebra. This was a difficult authoring decision based on fitting an acceptable symbol definition into the broader scope of notation, which includes that of both numerals and algebra. Each part has its separate chronology. Parts 1 and 2 are quasi-independent, but the reader should be aware that at early stages of development, both numerals and algebraic symbols progressed along chronologically entangled lines.

Definitions

sym•bol \ˈsimbəl\ n-s: Something that stands for or suggests something else by reason of relationship, association, convention, or accidental but not intentional resemblance.[1]

"Symbol" is a complex word. Webster's definition does not quite fit the collective experience of its use. For a more suitable fit to this book, we must extend the preceding definition to require a symbol to also be, or to have been, something cultural and nonarbitrary, something representative of an object or a concept that it does not resemble in sound or in look, and something that gives no preconception of the thing it resembles.

al•ge•bra \ˈal-jə-brə\ n-s: A branch of mathematics in which arithmetic relations are generalized and explored by using letter symbols to represent numbers, variable quantities, or other math entities (as vectors and matrices), the letter symbols being combined, esp. in forming equations in accordance with assigned rules.[2]

These days, the word "algebra" has a much broader meaning that spills into generalized rules of addition and multiplication and structural relationships between all sorts of mathematical objects. However, since this book is mostly about the symbols of pre–eighteenth-century algebra, Webster's definition is appropriate.

Note on the Illustrations

Primary sources illuminating the history of symbols are discussed throughout the book and, in some cases, represented as illustrations. Although print-quality scans were available for some of the original manuscripts illustrated in this book, for technical reasons it was necessary to typeset the textual figures on pages 69, 102, 141, 155, 156, and 158.

Part 1

Numerals

Significant Manuscripts and Initiators

BAKHSHÂLÎ MANUSCRIPT (date in dispute: 400–700). Indian.

Shows that the Indians had a place-value system in place before 700 AD.

BISHOP SEVERUS SEBOKHT (ca. 575–ca. 666). Syrian. Science and philosophy writer.

Author of the earliest known extant reference to Hindu-Arabic numerals outside of India.

BRAHMAGUPTA (598–668). Indian. Mathematician-astronomer.

His *Brahmasphutasiddhanta* (628) has the first known use of zero (a small black dot) as a number, not just as a placeholder.

HARUN AL-RASHID (8th century). Persian. Caliph.

Founded the House of Wisdom in Baghdad, a library and translation center that contained manuscripts of mathematics and philosophy translated into Arabic from many other languages.

AL-KHWĀRIZMĪ (ca. 780–ca. 850). Persian. Mathematician-astronomer-geographer.

Scholar in the House of Wisdom. Wrote the *Compendious Book on Calculation by Completion and Balancing (Algebra)*, 830 AD.

MAS'ÚDÌ (Abu'l-Hasan 'Ali) (ca. 896–956). Mesopotamian. Arab historian, adventurer.

His *Meadows of Gold* (957), a thirty-volume collection of histories of Persians, Hindus, Jews, Romans, and others gives a dependable tenth-century account of the nine Hindu-Arabic numerals.

GERBERT D'AURILLAC (Sylvester II) (946–1003) French. Pope.

Studied and taught mathematics and designed a counting board, called the Gerbertian abacus, with a place-value system using Roman numerals.

CODEX VIGILANUS MANUSCRIPT (ca. 976). Spanish.

An illuminated manuscript containing the first Arabic numerals in a Western manuscript.

RABBI ABRAHAM BEN EZRA (1089–1164). Spanish. Astronomer, mathematician.

His *Sefer ha-Ekhand* (*Book of the Unit*) described the Hindu-Arabic number symbols, and his *Sefer-ha-Mispar* (*Book of the Number*) described the place-value system and zero.

ROBERT OF CHESTER (12th century). English. Arabist.

Translated al-Khwārizmī's *Algebra* into Latin in ca. 1143, but was discovered only in the nineteenth century. Al-Khwārizmī's *Algebra* contains one of the earliest known introductions to the Hindu-Arabic numeral system.

JOHANNES HISPALENSIS (also John of Seville) (12th century). Spanish. Translator.

His *Arithmeticae practicae in libro algorithms* (*Book of Algorithms on Practical Arithmetic*) contains the earliest known Western description of Hindu-Arabic place-value notation.

LEONARDO PISANO BIGOLLO (Fibonacci) (ca. 1170–ca. 1250). Italian. Mathematician.

His *Liber abbaci* (1202) used the Hindu-Arabic system and was written in the vernacular for Italian tradesmen.

ALEXANDER DE VILLA DEI (ca. 1175–ca. 1240). French Minorite friar and poet.

His *Carmen de Algorismo*, written in Latin verse, explained the methods of computation involving Indian numerals including zero.

JOHANNES DE SACROBOSCO (1195–1256). English. Astronomer, monk.

His *Algorismus* was a European best-selling textbook on Hindu-Arabic numerals and how to use them in calculations.

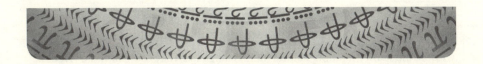

Chapter 1

Curious Beginnings

No one knows precisely when humans first began to deliberately leave marks for communication with others. Surely it was in that misty period of time, when herds of woolly mammoth freely wandered Europe, and all sorts of living creatures were following the northward spread of food and vegetation from the plains of Africa.[1] The ice of Europe had been receding for centuries in the slow ending of one of the great climate changes of all time. Most of the human population was still in southern Asia.[2]

That was between fifty and thirty thousand years ago, when humans had to think about day-to-day necessities for survival. Deep ontological thoughts—such as *where did I come from?* and *why do I exist at all?*—that could be shaped only by strength of language and metaphor potential were not likely. Even without a richly developed language, they must have had our natural urge to tell stories, the impulse to relay to others those pictures in the mind. These may have been fancies about thunderstorms, darkness, beasts, or even the puzzlement of dreams, but such is the nourishment needed to push language further.[3]

As language developed, so did contemplations of the experience of being alive. The twentieth-century preeminent folklorist Joseph Campbell told us that humans have always been "seeking an experience of being alive so that the life experiences that we have on the purely physical plane will have resonances within, that are those of our own innermost being and reality, so that we actually feel the rapture of being alive."[4]

Humans could have survived in their severe, menacing environments by some combination of instinct and intelligence, just as most mammals did and still do, without the spark exploding natural human language. They could have survived the freezing winters and scorching summers in an oral world without a written record, in a world without marks, signs, symbols, or paintings. Monkeys did; so did caribou.

What possessed those Neolithic cave painters to ignore the dangers of daily life while they sat, etched, scribbled, or painted? More than 40,000 years ago, the dwellers near the Cave of El Castillo in Spain bothered to stencil their hands against cave walls by blowing pigment.[5] For tens of thousands of years, humans had been leaving signification marks in their surroundings, gouges on trees, footprints in hard mud, scratches in skin, and even pigments on rocks.

A simple mark can represent a thought, indicate a plan, or record a historical event. Yet the most significant thing about human language and writing is that speakers and writers can produce a virtually infinite set of sounds, declarations, notions, and ideas from a finite set of marks and characters. Animals may have their languages, but they cannot produce an infinite number of communication signifiers from a finite number of sounds and gestures.[6]

From pigment-sketched mammoths on rocks to alphabets, writing developed through transitional stages. Pictures were clues to picture writings, which in turn were clues to ideograms, and so on through modifications, all the way to early metaphorical poetry and modern writing. A "pictogram" is a picture that resembles what it intends to mean. In Asia, such writing became the foundation for modern Chinese hanzi and Japanese kanji. In today's world, a picture of a knife and fork might represent a restaurant. A slash across the knife and fork would be an "ideogram": it might signify *no eating allowed*. Whereas a pictogram depicts objects, an ideogram expresses meaning through similarity or analogy. To signify the word "home" by ideogram, for example, in early domestic China, you would combine the pictogram for "roof" with the pictogram for "pig" to make the word "home." For at least thirty thousand years, stories have been told through pictures, and as the years progressed, the stories became more elaborate.

Some years ago, a friend returned home from Thailand and gave me a gift of a Hmong embroidered "story cloth" that was bought from a weaver living in a refugee camp. It depicted the story of daily life during the Vietnam War. From it one could "read" the cycle of life. There is the birth of a child, work in the fields, falling in love, a wedding, and a new birth—a whole story without a written word.

Pictograms are easy to understand in a simple world of simple stories. The problem comes when the storytelling is more complex. Imagine the *Odyssey* "written" in pictogram characters. Who would fully understand it? It would be too elaborate and too laborious to absorb, and quite possibly too inflexible for the metaphoric complexities of serious poetry. Far better is to have the characters represent the phonemes of speech, so one utterance is distinguished from another—*a* for "ah," *b* for "be," and so on.

It's one thing to have words and quite another to think about the words themselves. Writing sentences is altogether different from talking; it must have come after a great deal of social growth, after the first civilizations, after kings and emperors, and long after adventurous tribesmen started wandering beyond the familiar for adventure and trade.

If you ask a person in the street what he or she thinks is the most important invention in the history of civilization, you are likely to be given the proverbial answer: the wheel. Surprisingly, the wheel didn't come into existence before the late Neolithic Age and possibly as late as the early Bronze Age. That would put it somewhere between 6,000 and 3,500 BC. The earliest depiction of a wagon with wheels can be seen on a ceramic pot that was excavated in Bronocice, Poland, in 1976. The Bronocice pot dates back to ca. 3500–3350 BC.[7] But with the new agriculture of that period, the wheel should have been an obvious invention—after all there was circular pottery, and slices of tree trunk must have given good clues to the enormous utility advantages of a rolling disk. Surely rolling logs were used before the true simplicity of the wheel. But the wheel is not just a rolling disk. It involves the relatively complex concept of wheel and axle, combined.

What about the alphabet? It is surely a contender. I would argue that most of the other significant inventions that have made our lives possible could not effectively exist without the alphabet, or at least some other clever way of writing the words we speak. True, that person on the street might argue that the great pyramids of Egypt could not have been built without the wheel in the form of rolling logs to help the slaves, and that the tall stone buildings of the world could not have been built without the wheel and axle. The wheel would have come to the world sooner or later, but some form of writing the sounds we make trumps all.

Modern alphabetic writing is a rough mimic of spoken language. Before any evidence of an alphabet, there was Sumerian picture writing, where each syllable of the Sumerian language was a distinct picture impressed by a wedge stylus in clay. Originally, the impressions were meant to be pictures of objects with the same syllabic sound of the word that was to be conveyed. This was a different sort of writing than that of mere pictographs. Fortunately, spoken Sumerian was a language of words made from many syllables, and often the syllables themselves were the names of concrete objects. The writing consisted of marks, each denoting a syllable. For example, a picture of a house being held in a hand might signify "household."

Hieroglyphic picture writing was used in the Mediterranean area around Egypt at roughly the same time as Sumerian picture writing, which went through several transitional stages before slipping away from its pictorial character to evolve into a pure sound-sign system, and eventually something alphabetic.[8] By the time the Phoenician alphabet was introduced, sometime before the first millennium BC, numerous cultures in almost every part of the world had developed some form of representational writing using pictorial symbols. This gave those cultures the means of immediate communication and a means of leaving a record of knowledge for future generations as well.

Unlike the phonetic writing that we have today, in which the symbols of each word represents the sounds of the spoken words, pictorial writing was an indicator of the meaning of the spoken word, not the sound. By the middle of the first millennium BC, however, pictorial writing was replaced by phonetics.

Pictures can be used to represent words through their sounds. In English, for example, you could write "I believe" by juxtaposing the pictures of an eye, a bee, and a leaf.[9] Meaning in hieroglyphics was represented through context, just as it is in phonetic writing. Phonetic writing, however, has at least one important advantage over pictorial: it can express far more combinations of thoughts and ideas. One might also argue that writers can work in a much freer playing field to invent richer metaphors.

It is not surprising that the need to write came from the need to record memories, not stories. The earliest documents are of accounts, names, recipes, and itineraries. As the skill of writing spread, so did the reasons. One can imagine graffiti on public buildings, secret notes and magic formulas passed to other people, writings to help one's memory, or epitaphs for one's tombstone. Such memories and epitaphs "call men and women to a deeper awareness of the very act of living itself, and they guide us through trials and traumas from birth to death."[10]

At first, writing was limited to the initiated, mostly the priestly sects or special classes who were trained; once it settled to some standards, however, its power had profound effects on spoken language. Educated peoples from distant lands and roughly similar languages were soon able to share a common written language, thus fixing the verbal traditions and creating a common bond of experiences between foreign lands and separated times.

The beginnings of civilizations and cities coincide strikingly with the construction of temples and the rise of priestly classes, which attracted bright recruits from the common populace. Primitive agrarian life slowly included a temple life with priest kings who built their empires. This may have been a result of the growth of agrarian cultures, which depended on calendars that were understood by the priests and held by the temples for seasonal rituals. Thus priests, human representatives to the gods, governed the earliest civilizations.[11] Their temples were observatories, libraries, clinics, museums, and treasure houses. Though the Babylonians had relatively extensive star catalogues by 1200 BC, it was the Egyptian priests who—believing the sky divine—mapped out the stars and constellations as early as 3000 BC.[12] The complexities of star map calculations, along with land surveys and taxes, required writing numbers beyond the simple low numbers that were useful in accounting for sheep in the fields.

Primitive humans had simple needs. At first, counting was limited to very low numbers. The shepherd could know that a sheep was missing from the flock without needing to count. Any ape could do that—that is, know that a member of the family is missing. To know that something is missing is a qualitative, rather than quantitative, notion of sets. Facets of primitive life didn't require any real sense of number. No one needed to know what number *is*.

Yet still, for some wonderful reason that seems almost inexplicable, humans—even primitive humans—have always had an uncanny ability to recognize numbers beyond the values for which they had words. Children today are taught to recite numbers in preschool to get a sense of the words associated with quantity. They can easily recite the numbers from 1 to 10. Reciting numbers, however, is not the same as understanding what those numbers actually mean. A three-year-old may be able to count to 5 without understanding the one-to-one correspondence between the words "one," "two," "three," "four," "five," and the five fingers on one hand. That correspondence, whenever it occurs in child or human development, is a gargantuan leap of cerebral maturity. We don't notice the moment of that leap. There doesn't seem to be any "aha!" experience at that moment. Having five fingers on each hand does not seem to naturally suggest a one-to-one correspondence with the first ten numbers. Until the middle of the last century, several aboriginal tribes in Australia had no words for numbers, but could count by making marks in the sand.[13] Curiously, there were—at least before the last century—several indigenous tribes of Australia, the Pacific islands, and the Americas that had no words for numbers beyond four, suggesting that the modern concept of numbers as one-to-one counters had not yet matured.[14]

In both the East and the West, mathematical writing predates literature by more than a thousand years. It even predates the oldest surviving written story, *The Epic of Gilgamesh*, a Sumerian poem that was written more than a thousand years before the *Iliad*. We have no direct evidence as to where or when numerical writing first occurred, just as we have no direct evidence as to where or when writing first began to develop. Some would attribute the earliest concepts of numerical writing to the Chinese, as far back as the Early Stone Age. That seems doubtful. But it appears reasonable that it coincides more or less with cuneiform Sumerian number writing dating back to 3400 BC.[15]

Like the art found in the caves of southern France and northwestern Spain, number writing came about through the human endeavor to record.[16] One of the world's oldest extant written records (German Archeological Institute Museum number W 19408,76+) seems to be an exercise in calculating the areas of two fields, written sometime in the late fourth millennium BC. It is a collection of fragmented clay tablets found among the reused building rubble of the city of Uruk. Its carbon date (ca. 3350–3200 BC) predates any known evidence of writing, at least of writing that we agree is phonetically associated with a spoken language.

Traces of Sumerian number writing on clay tablets with numbers as large as 10,000 have been found in caves from Europe to Asia. Egyptian hieroglyphics had a distinct symbol for the number 10,000. By 1600 BC, the algebra problems in the famous Rhind (or Ahmes) papyrus presented simple equations without any symbols other than those used to indicate numbers.

Chapter 2

Certain Ancient Number Systems

Call them what you wish—Babylonians, Sumerians, or Akkadians. We have heard their stories before. Almost every history of early Western mathematics begins with the Babylonian conception of number, a so-called sexagesimal (base 60) system for writing large numbers, formulations of multiplication tables, and ideas for astronomy. But who were those Babylonians, and why were they the ones to first come up with human civilization, culture, art, and science?

To answer, examine the geographical region of the Fertile Crescent, that crescent-shaped region between the Eastern Mediterranean and the Persian Gulf, and running through southeastern Turkey to Upper Egypt. It happens to be a unique area responsible for the spread of wild emmer wheat, wild einkorn, and wild barley, and therefore an exceedingly favorable area for the birth of local agriculture.[1] Within the Fertile Crescent lies an area near the Tigris-Euphrates valley. Generally, the term "Babylonian" refers to things related to far more than just the city of Babylon, and essentially to a wide geographic area that today includes southern Iraq, Kuwait, and parts of western Iran. It is an area near and between two great rivers that converge close to modern Baghdad, then diverge and zigzag until they meet at Al Basrah in southern Iraq, just north of Kuwait, before pouring into the Persian Gulf. If you look at a map of these great rivers, you cannot avoid being impressed by their meanderings. The Tigris wanders south of Baghdad as if it were a water snake that cannot make up its mind whether to go southwest or northeast. In some places—near Suwayrah, for instance (figure 2.1)—it can take two hours to navigate the Tigris by boat only to find that a ten-minute walk over land will bring you to the same spot. In other places, a half-hour walk will bring you to the same spot it would take a boat six hours to reach. This means that the land between two relatively long lengths of the river may be easily irrigated. Even today much of the banks of the Tigris is undeveloped farmland. There are few long sharply twisting rivers in the western world. Rivers generally go places from high elevation to lower. There are rivers in northern Europe that have sharp meanderings—for instance, the thousand-kilometer Elbe—but northern climates were not terribly welcoming to winter crops. Though the Tigris-Euphrates valley terrain was not ideal for farming, the great rivers, with their many tributaries and canals running slowly through, were outstanding for irrigation. Small villages grew along the rivers that cut through the moderately flat countryside south of Baghdad to collectively become the first urban centers in the West. Many of the ancient tributaries and canals along the alluvial plain south of Baghdad that were around when the first settlements carpeted the region are now gone, dry.

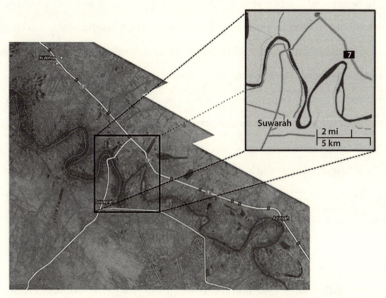

FIGURE 2.1 Section of the Tigris River near Suwayrah. Source: Google Maps.

If you lived in King Hammurabi's time, however—that would have been almost thirty-seven hundred years ago—and wanted to settle down to plant crops for your family, what better place could you find? Southern Mesopotamia. Its flat marshlands, vast tracts of fertile soil, and abundant wildlife were ideal for growing barley and managing sheep and goats. There is where the first urban civilizations were anchored.

Southern Mesopotamia leaped ahead of other regions in urbanization and efficient farm irrigation near capital cities along rivers and canals of Kish, Nippur, Lagash, Uruk, Eridu, Shuruppak, and Ur. Babylon was at the center of an empire extending far beyond southern Mesopotamia into the northwestern bend of the Euphrates River.

Snaking rivers and a prolonged farming season helped, but there must have been something else that made that place so special. Was it the soil, the trade route, or pedigree of its ancestral lineage? According to Bill Arnold, the author of *Who Were the Babylonians?*, it was neither the soil nor the trade route.[2] Egypt, he argues, "was largely isolated from the rest of western Asia because it was limited to the narrow band of hospitable land created by the Nile Valley," and so was a land of few invasions, a land limited to very few cultural diversities. Mesopotamia, on the other hand, was vulnerable at almost all its borders with a continuous infusion of distinct nationalities bringing the usual riches of diverse cultural influences.

Babylonia was a "melting pot" of antiquity, not through any hospitality to foreigners, but rather through its lack of natural landscape barriers and thus its vulnerability to foreign invasion. The open plains of the south and the waters of the gulf were easy entrances, and the hills of the east and northeast were easy passageways to the urban centers of Babylonia. Frequent invasions by seminomadic populations collaged the whole region into contrasting ethnicities that continuously mingled and fused.

Southern Mesopotamia's large and growing urban centers were sustained by a unprecedentedly wide socioeconomical linkage that, for the first time in history, required a managerial workforce to administer, organize, and account for trade and labor. That's where records had to be kept. That's where recorded accounts began. Written in clay, the accounts were collections of symbols alongside pictogram descriptions of the objects being accounted—land, people, livestock.[3]

In the early 1900s, when the Ottoman Empire was crumbling and barely regulating trade in minor antiquities (other than with bribes and bureaucratic obstructions), the American diplomat, antiquary collector, novelist, and itinerant archaeologist Edgar James Banks bought hundreds of cuneiform tablets on the open market. He later transported those tablets to America and sold many to museums, libraries, and collectors. There was one tablet in his collection that came to be of special interest to mathematics historians. It was found at Senkereh, an archaeological site near the ancient Babylonian cities of Larsa and Ur, the birthplace of Abraham in southern Iraq.

Banks sold it to the New York publisher George Arthur Plimpton for $10 in 1922 (approximately $130 in today's value, according to the American consumer price index).[4] It is always difficult to reconstruct pieces of a culture from fragments of history, and the story of Plimpton 322 has many sides.[5] In 1945, the mathematical historians Otto Neugebauer and Abraham Sachs interpreted it as containing a list of Pythagorean triples—that is, a list of integer solutions to the equation $a^2 + b^2 = c^2$. What makes this striking is that it predates the Western idea of a Pythagorean theorem by more than a thousand years, and yet suggests that the Babylonians must have had a suspicion of some sort of Pythagorean theorem. Recently, however, the mathematical historian Eleanor Robson, working at the University of Cambridge, has given a strong case for interpreting the tablet to be a teacher's aid, designed to generate problems about right triangles, not at all a proto-Pythagorean theorem.[6]

The illustration in figure 2.2 is a pen-and-ink sketch of a Babylonian tablet made in the ancient city of Nippur at the center of Babylonia about 3,700 years ago. The marks are not tiny bird footprints, but rather impressions made from a wedge-shaped stylus. When pressed into moist slabs of clay, the stylus would leave a print of either the form Y or the form $\mathsf{\prec}$.[7] The slabs would then be baked.

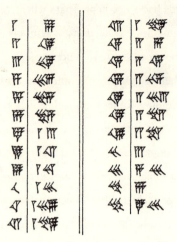

FIGURE 2.2 The Nippur Tablet. 𒁹 stands for 1 and 𒌋 for 10. Redrawn from R. Creighton Buck, "Sherlock Holmes in Babylon," *American Mathematical Monthly*, vol. 87, no. 5 (1980): 335–345. Reprinted with permission of the Mathematical Association of America.

Examine the left column, reading from top to bottom. Knowing nothing of the ancient script, we can guess that the column represents the numbers from 1 to 12. What about the second column from the left? If our first guess is correct (and how could it not be?), we would know that the first symbol in that column represents the number 9. What could the next number down be? It must be a juxtaposition of the symbol for 10 and the symbol for 8. Could it be 18? By the same reasoning, the third symbol seems to represent 27. Hmm … could the second column be the multiples of 9? That seems to be true until the sixth line down—that is, $6 \times 9 = 54$. At the seventh line down, something strange seems to happen. The symbol looks as if it is a 4. But is it? If it is, then that second column is not a list of multiples of 9. So what could it be?

We notice that there is a space between the first wedge mark and the other three. If there is a hope that the second column is a list of multiples of 9, then the seventh symbol should be 63. Perhaps the space indicates that we should multiply by 60 before adding the three wedges. That would give the correct multiple of 9.

Testing this hypothesis with the remaining entries, we find that our reasoning works:

$$8 \times 9 = 1 \times 60 + 12 = 72$$

$$9 \times 9 = 1 \times 60 + 21 = 81$$

$$10 \times 9 = 1 \times 60 + 30 = 90$$

$$11 \times 9 = 1 \times 60 + 39 = 99$$

$$12 \times 9 = 1 \times 60 + 48 = 108$$

And it holds for the next two columns. So we have here an example of a very early clever use of notation, where a "space" is used as a symbol.

Our number symbols—that is, our current ones—are quite different and far more sophisticated. The number 72 represents 7 times 10 plus 2. We need symbols for just ten numbers (0, 1, 2, 3, 4, 5, 6, 7, 8, 9) in order to represent any number we wish. The Babylonian system needed just two symbols, though, to the uninitiated eye, it may seem as if it needed fifty-nine distinct marks. To designate a number less than 60, symbols of smaller numbers were systematically clumped together, one almost touching another. For example, the number 39 would be written as

We write sixty-one as 61 and know that we don't mean 3,601. How did the Babylonians distinguish 61 from 3,601? The number 61 is represented by ୮ ୮, and the number 3,601 is represented as ୮ ୮. The only difference is the number of blank spaces separating the wedge marks. (୮ ୮ sandwiches one blank space, while ୮ ୮ contains two.) However, since a blank space has no visible boundaries, it is difficult to know (especially when those marks are done by hand) how many blank spaces separate the marks. There is a problem. A blank is simply a vacancy, and two blanks might still look like just one vacancy.

You might think that context had to play some role in distinguishing numbers by the sizes of their relative values, just as it does in the ambiguous circumstances of language. For instance, distinguishing ୮ goats from ୮ goat is easy: the first has the hint of plurality, and therefore must mean 60 goats, while the second is singular and must mean 1 goat. Would context distinguish ୮ ୮ (3,601) from ୮ ୮ (216,001)? Possibly.

Someone had to come along and devise a tool to make the system work. From the benefit of twenty-first-century retrospection, we see clearly what that tool is. Let any doodle stand in for a blank space—say, 𝄢. Then the reader could easily distinguish between ⌐𝄢⌐ and ⌐𝄢𝄢⌐. So why was that not done?

It was. Yet, like Rome, and unlike modern Beijing, it did not suddenly appear one day. Someone had to come up with a clever plan. Indeed, it had taken more than a thousand years for that to happen. At some time between 700 and 300 BC, someone thought of using a symbol looking very much like 𝄢 to represent a blank space. It was the invention of a placeholder, the Babylonian zero, though not the proper notion of a modern zero. It was then possible to distinguish ⌐ ⌐ from ⌐ ⌐ without having to rely on context alone.

As odd as this system must seem to us, it was brilliant. As long as one could distinguish spaces, a Babylonian arithmetician could write numbers of any value using just two symbols, and that exotic mark to demark a space.[8]

Long before Babylonian scribes pressed reeds in clay in the hot sunshine of Nippur, Egyptians were engraving hieroglyphics into stone, metal, and wooden monuments. That was when numbers were simply pictures of objects, when each power of ten had a different symbol. The number 1 was represented by a vertical staff; 10 by a staff bent in the form of a semicircle; 100 by a snail figure; 1,000 by a lotus plant; 10,000 by a pointing finger; 100,000 by something that looks like a small bird, or possibly a fish; 1,000,000 by a man with his hands raised, as if bewildered by the enormity of the size (figure 2.3).[9]

Some Egyptologists have speculated that the snail figure is a coiled rope, that the small bird is really a frog, and that the man who seems to be worthy of representing a numeral as large as one million is really a god. And, since Egyptians had no need for any number beyond, say, ten million, that god also represented any amount that was a very great amount. Number writing was additive with larger representatives always placed to the left of the smaller. So, for instance, the number 3,601 would be written as (from right to left) a picture of one vertical bar, six snails, and three lotus plants. The scheme works nicely, with no need for placeholders, a distinct advantage over the Babylonian. The number 3,601 is well distinguished from 36,001. To write 36,001, there would be no ropes, and no lotus plants, just a bar, six lotus plants, and three fingers. There is no confusion between 61 and 3,601. And moreover, there is no confusion between 36 and 3,600, since the latter would be a display of six ropes followed by three lotuses.

Early Egyptian number writing was an additive system. To write 1,005, one would simply join a lotus plant against five vertical staffs; however, no more than four staffs would be adjoined, so the scribe would divide the staffs into two groups. Sometime after the second millennium BC, a multiplicative system came into being. To write the number 2,000,000, a scribe writing on papyrus would draw a man on top of two staffs. But there are a great many puzzling questions that Egyptologists still cannot answer. For example, in hieroglyphics the unit fraction $\frac{1}{2}$ was written as the picture ⊏ and $\frac{2}{3}$ was written as ⊕.

ǀ	∩	℮	𝕀	𝔩	↘	👤
1	10	100	1,000	10,000	100,000	1,000,000

FIGURE 2.3. Early Egyptian number writing. From Florian Cajori, *A History of Mathematical Notations* (New York: Dover, 1993), 12.

The Hebrews had a different scheme. Their alphabet has twenty-two letters, each symbolizing a number (see table 2.1). There are five more letters that are used only at the ends of words. They are ך, ן, ם, ף, and ץ representing 500, 600, 700, 800, and 900.

To represent thousands, one would start from the beginning and place two dots above the letter. So אָּ would represent 1,000; בּ would represent 2,000; and so on. Now here is the tricky thing. Hebrew is read from right to left, and numbers above a thousand could be written two ways. As with all the Hebrew number schemes discussed on these pages, each culture's number scheme went though many trials and changes over the centuries. By the eighth century AD, the two-letter symbol הא would mean 5,001. The letter א would ordinarily represent the number 1, but when it appeared to the right of another letter—say, ה—it would represent 1,000. There would be no confusion, because, even though the letters were read from right to left, their number equivalents were understood to descend in value. Hence the ה to the left of the א would signify 1,005.

Table 2.1. THE HEBREW ALPHABET

א	(Aleph)	1	ל	(Lamed)	30	בַּד
ב	(Bet)	2	מ	(Mem)	40	
ג	(Ghimel)	3	נ	(Nun)	50	
ד	(Dalet)	4	ס	(Samekh)	60	
ה	(He)	5	ע	(Ayen)	70	
ו	(Vav)	6	פ	(Pe)	80	
ז	(Zayin)	7	צ	(Tsadi)	90	
ח	(Het)	8	ק	(Qof)	100	
ט	(Tet)	9	ר	(Resh)	200	
י	(Yod)	10	ש	(Shin)	300	
כ	(Kaf)	20	ת	(Tav)	400	

This scheme works nicely. The number 9,686 would be written טחוו. Notice that the letter ו appears twice, and yet it is considered two different numbers. Standing alone, it represents a 6. Reading right to left, the first ו must have a value between that of the ט and the ח, and therefore must mean a number between 9,000 and 80, and hence that ו must be 600. The ו in the last position must mean the smallest number it could possibly represent, which is a 6.

It is the nature of symbols in general to connect unrelated meanings in order to create a state of mind. In Hebrew, the number 15 would be naturally written from right to left as י (the symbol for 10) plus ה (the symbol for 5). Writing 15 in that way, however, would also be writing the first two letters of the name of God. So the number 15 was (and still is) written as 6 + 9 (טו) instead.

The Greeks borrowed the Hebrew system for representing numbers. They too had each number represented by a letter of their alphabet—a terribly inconvenient scheme for representing large numbers.

α β γ δ ...

1 2 3 4 ...

Why didn't they adopt the genius of the Babylonian system, with its placehold-ers and relative ease of writing large numbers? The Babylonians had the right idea of positional notation, the clever idea of using the same digits to represent multiples of different powers of 60. How could those mathematically resourceful Greeks miss such an inspired idea? With all that they did—their organization of logical thought, evidence, and proof; their understanding of geometry and the irrational; their pow-ers in resolving issues of number theory through geometry—why did they not see a better way of handling numbers to make arithmetic easier? One answer might be that some form of abacus was used for most calculations.

Perhaps it was because their interest was to grasp the grand scope of mathemat-ics itself. Calculation was not really their game, though surely there were a great many mathematicians doing nondeductive mathematics as well. Theirs was the de-velopment of a strictly deductive science, proof, solutions, universals, perfection, and an understanding of Euclidean space and the relations of objects that fill that space, all done with a relatively awkward number system, and a level of sophistica-tion that hardly needed a number system at all!

Sometime during the eighth century BC, the Greeks adopted the Phoenician al-phabet, and with it "acrophonic" numerals, symbols derived from the first letters of the written words representing numbers (see table 2.2).

It was a slow process leading to other systems competing for popularity during the fifth century BC, when more global commerce took effect with the Hebrews, Syr-ians, and Phoenicians, who had their own alphabets. As it happened, those alphabet-clever Phoenicians lifted Egyptian hieroglyphic signs, gave them unique sounds, and represented those sounds as letters, yet oddly did not use their own alphabet to rep-resent numbers—rather, they used a system of vertical bars.

Table 2.2. THE GREEK ACROPHONIC SYSTEM

1	I	Ἰῶτα
5	Π	Πέντε
10	Δ	Δέκα
100	H	Ἡκατόν
1,000	X	Ξίλιοι/χιλιάς
10,000	M	Μύριον

Note: To write 5 times a number, place the number under the umbrella symbol ⌐. For example H̄| signifies 500.

Sometime after the fourth century BC, the Greek sequential alphabetic number system won the competition and displaced the old acrophonic system. Like the Hebrew system, the Greek alphabetic system became the standard.[10]

It was quite awkward when describing large numbers. Even Archimedes, when writing his ingenious *Sand Reckoner*, on estimating the number of grains of sand that it would take to fill the universe, resorted to using words, not notation, to describe such large numbers. His answer, in our convenient notation, was close to 10^{51}, well below[11] the more correct answer of about 10^{90}.

But why did the sequential alphabetic number system win? Why did the Greeks abandon the acrophonic system for the alphabetic? Could it have been simply to have shorter representations? The number 1,884 in acrophonic notation is χ H̄|HHH Δ̄|ΔΔΔIIII. In alphabetic notation, it becomes αωπδ. Wouldn't it be harder to remember the numerical values of twenty-seven symbols than to remember just six? Yes, but over time, a schoolchild could memorize the values, just as well as memorizing the order of the letters.

No. It seems that there is a conceptual difference. Florian Cajori, the early-twentieth-century mathematics historian, in his consummate history of mathematical notation, considered the following two arithmetic identities (keep in mind that the plus and equal symbols did not appear before the sixteenth century):

$$HHHH + HH = \overline{H}|H,$$

$$\Delta\Delta\Delta\Delta + \Delta\Delta = \overline{\Delta}|\Delta.$$

The corresponding equalities written in alphabetic representation are

$$\upsilon + \varsigma = \chi,$$
$$\mu + \kappa = \xi.$$

There was a choice. Both representations were human-made and both competed for the favors of arithmeticians and scribes who had to use them. Both were cumbersome. Only one won out. Why didn't the Greeks, who surely knew about the Babylonian system, come up with a more clever system that uses placeholders? Why was it left up to the Indians east of the Punjab to come up with the smartest system of all?

We can see the shadowy hints of a place-value system in figure 2.4. The first ten Greek letters represented the first ten numbers. To represent numbers from 11 to 19, one would write ια, ιβ, ιγ, ιδ, ιε, ιϛ, ιζ, ιη, ιϑ; which would have meant 10 + 1, 10 + 2, and so on. Then, to represent 20 to 29, one would write κ, κα, κβ, κγ, κδ, κε, κϛ, κζ, κη, κϑ; which would have meant 20, 20 + 1, 20 + 2, and so on.[12] Though there were some invented symbols, such as the strange symbols for 90 and 900, the creators should have realized that the placing of a symbol could dictate its value. To write the number 23 in the Greek system, where place does not dictate value, one is forced to introduce a new symbol (κ) to represent 20. To write the number 23 in a system where place does dictate value, all that is needed is β, the symbol that represents 2, a symbol already well defined. When β appears in the second place it means 20, not 2. The number 23 could have been written as βγ.[13] I bring this up here to show that a choice of symbol notation can be an obstruction to future advancement. As with any of the other alphabet numeral systems, the Greek scheme was fine for low numbers, but awkward for large numbers.

These ancient alphabets were not just collections of concrete linguistic elements with individual identities, but building blocks ripe for multiple meanings. Fifth century BC Greeks believed that everything in the world could be connected to whole numbers. The number 2 (the letter "β") meant opinion, 3 ("γ") harmony, and 4 ("δ") justice. Odd numbers were male; even numbers female. The number 5 ("ε") symbolized marriage, possibly because it was the sum of the first even number with the first odd number. And the number 10 ("κ") was holy because it was the sum of the first four dimensions (point, line, triangle, tetrahedron), $1 + 2 + 3 + 4 = 10$. So we begin to see that all sorts of metaphorical states of mind are emboldened by these ancient number systems.

α	β	γ	δ	ε	ϛ	ζ	η	θ	ι	κ	λ	μ	ν	ξ	ο	π	ϟ
1	2	3	4	5	6	7	8	9	10	20	30	40	50	60	70	80	90

ρ	σ	τ	υ	φ	χ	ψ	ω	ϡ	,α	,β	,γ,
100	200	300	400	500	600	700	800	900	1,000	2,000	3,000

etc.

M	$\overset{\beta}{M}$	$\overset{\gamma}{M,}$	etc.
10,000	20,000	30,000	

FIGURE 2.4. Greek sequential alphabetic system. The letter representing 6 is ϛ, a cursive *digamma*, an ancient letter that disappeared from the Greek alphabet sometime before the seventh century BC. It looks like the sigma used at the end of words, but its sound is very different. It is important to keep this in mind when we talk about *the alternate sigma* as a nonnumeral symbol in part 2. Note that 6 is represented by ϛ, even though the true alphanumeric order should have ζ represent 6.

The Roman system, which was closely related to the Greek acrophonic system, used the principle of addition to display large numbers along with a clever idea of a subtraction rule: when a smaller number was placed to the left of a larger, it meant subtract the smaller from the larger (see table 2.3).

Table 2.3. LATE ROMAN NUMBER SYMBOLS

1	I
5	V
10	X
50	L
100	C
1000	M

Hence, 83 could be written as XXCIII instead of the longer version, LXXXIII. Good number grammar required numbers to be displayed by their shortest possible length, but it was not always applied. There were variations. For example, by the fourth century AD, we find places where writers used a horizontal line above a number to indicate a thousand times that number, so $\overline{X}$ would mean 10,000, rather than 10. Vertical lines to the left and right of a number with a bar over it would indicate one hundred thousand times that number, so $|\overline{X}|$ would indicate 1,000,000. Like the Greek scheme, it too was terribly inconvenient for representing large numbers.[14] We still use Roman numerals for dates, though I wonder why.

Though the Aztec numerals have no direct historical connection to those of Asia, Africa, and Europe, they show similarities. Aztec numerals began with dots for units up to 9. After 9, they became pictorial. A full feather was considered to be 400, and so a quarter feather was 100, a half feather was 200, and three-quarters was 300. The symbol for 8,000 was a purse that presumably contained 20 times 400, though the purse itself had no clear indication of that product. (See figures 2.5 and 2.6.)

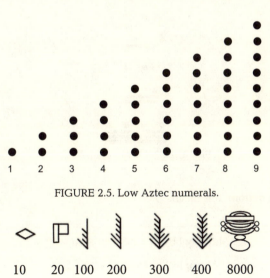

FIGURE 2.5. Low Aztec numerals.

FIGURE 2.6. Higher Aztec numerals.

As with other systems on other continents, the system was additive. Unlike the Babylonian single base system, however, the Aztec system had three bases: 20, 400, and 8,000. To write, say, 26,504, an Aztec would write what is shown in figure 2.7.

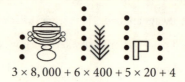

$$3 \times 8,000 + 6 \times 400 + 5 \times 20 + 4$$

FIGURE 2.7. Writing 26,504 in Aztec.

The Mayan (date unknown, yet likely in the Classical Period, 250–900 AD) approach was close to being a vigesimal (base 20) system. This Mayan arithmetic system is pre-Columbian, and yet the notion of adding and carrying characters is similar on two continents that had no human contact for over 50,000 years. Similar to the Babylonian system, it made use of the placeholder utility of zero in a system of dots, bars, and columns. A dot represented a unit, and a bar represented five units. For example, 3,212,199 would be a written as shown in figure 2.8.

FIGURE 2.8. Mayan numerals.

From top to bottom, this means

Multiply 1 by 18 × 20 × 20 × 20 × 20

Multiply 2 by 18 × 20 × 20 × 20

Multiply 6 by 18 × 20 × 20

Multiply 2 by 18 × 20

Multiply 13 by 20

Units level (19)

The sum, from bottom to top is

$$19 + 260 + 720 + 43,200 + 288,000 + 2,880,000 = 3,212,199.$$

The system does make arithmetic easy. To add two numbers, write them as columns, add their respective rows, and then add the resulting column, using a carryover, as we do in our modern system of addition. For example, to add 55 to 151, a Mayan would write:

and then add the figures in each row to get:

Four bars in the lower figure would "carry up" to become one dot of the next level:

Although the number 206 is easily written, a lower number such as 20 is more problematic. Mayans could not just place a dot at the second level when there was nothing in the units level. So they cleverly designed a symbol that would stand for the empty level, a zero that looked something like ⬬. The number 20 could then be written as ⬬.

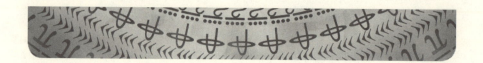

Chapter 3

Silk and Royal Roads

Master Sun says: [The function of mathematics] is to investigate
the assembling and dispersing of the various orders [in nature], to
examine the rise and fall of the two *qi* (i.e., yin and yang).
—*The Nine Chapters*[1]

Topography and frequent pounding of hooves and shoes formed the East–West
route connecting China to India and India to Persia. There were no road crews.
The Silk Road was not one particular road, but rather a series of land and sea routes
crisscrossing Eurasia, passing over 4,000 miles of wildly rugged terrain, and connecting to other routes traveled mostly by Indian merchants, agents, and explorers.
Formed sometime around the second century BC, it connected to the Royal Road
in the Zagros Mountains of Persia, where postal offices relayed mail and where one
could find fresh horses for a journey all the way to the Mediterranean. Satin, silk,
hemp, perfume, spices, jewels, glass, and medicines traveled westward; gold, silver,
carpets, and wines eastward. As with all major commercial trade routes, the Silk
and Royal Roads were also communication lines between cultures, religions, and
philosophies as well as germ itineraries for minor and serious illnesses.

Teaching and knowledge of philosophy, science, and mathematics also passed
along those intercountry highways. Commercial trade was done mostly through
bartering, but fair bartering required at least a rough estimate of value, an understanding of conversions of weights and measures: square areas of silk, or weights of
gold, or value of coin. An Indian who traded with both Persian and Chinese agents
had to understand the mathematics of commerce and be able to convey and under-

stand some kind of numerical information, possibly through converting between labels of Western and Eastern numerals.

So much of the history of Chinese mathematics has been lost or destroyed over the centuries, largely through book burnings ordered by despotic emperors, that we in the West tend to believe the myth of Western dominance of mathematical origins. The oldest recorded Chinese mathematics, including the first written numerals, dates back to the Shang dynasty (1600–1029 BC). In 1899, archaeologists excavated thousands of bones and tortoise undershells from a site that was once the capital of the Shang dynasty at Xiaotum in south-central China. Since then, tens of thousands of new findings have been collected and studied. They have numerical symbols marking numbers of enemy captured or killed in battle, numbers of birds and animals hunted, numbers of animals sacrificed, and tallies of many other achievements.[2]

By the beginning of the Early Han dynasty (206 BC–9 AD), Chinese numerical characters were established as a decimal system looking very much like the Chinese numerals that are used today (see figure 3.1).

1	2	3	4	5	6	7	8	9	10
一	二	三	四	五	六	七	八	九	十

10^2	10^3	10^4
百	千	萬

FIGURE 3.1.

The number 26,999, for instance, would be written as

二萬六千九百九十九 .

We have to celebrate just how clever this is. From left to right, it reads: 2 ten thousands, 6 thousands, 9 hundreds, 9 tens, and 9. It is, by all means, a decimal system. Why so clever? The zero is never needed, at least not as placeholder. To write 20,009, you would simply write

二萬九 .

We really must admire the Chinese for this. When there are no tens, hundreds, or thousands, just don't include them. No need for a zero!

I expect it would be every mathematics historian's joy to know who those smart Chinese mathematicians were who came up with such brilliantly simple ideas that include not only a slick system for writing numerals for commerce but also clever ideas for land surveying and astronomy. Alas, because of wars, book burning, and destruction of manuscripts, almost nothing is known about even the most major creative contributors.

This ingenious numeral system gave the Chinese a way to "name" large numbers, but something else was needed to do practical arithmetic. Once again, the Chinese had a terrific system: counting rods.

Long before the first millennium, the Chinese commonly used counting rods made from animal bones or bamboo to represent numerals one through nine (figure 3.2) in a base ten positional number system.[3] It was a arrangement of vertical and horizontal bars in a decimal system very like our own, except there was still no notion of a zero placeholder.

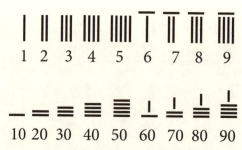

FIGURE 3.2. Chinese counting rods.

Naming numbers and finger counting may be fine for representing small quantities, but addition, multiplication, and division require some moving and removing: writing, scratching out, and rewriting. In the absence of cheap paper in the first century BC, counting rods, which could be quickly moved and removed in the course of a sequence of calculations, were most effective. Like the Hindu-Arabic system, the Chinese written numerals and the counting rod numerals were positional and easy to use, not just for representing numbers, but also for facilitating computation and mathematical concepts.

Knowledge of how to use counting rods was assumed in a Han dynasty (206 BC–208 AD) manuscript *The Nine Chapters on the Mathematical Art* (*Jiuzhang suanshu*),[4] an immense collaborative work of 246 problems written on traditional connected bamboo strips.[5] Historians believe it to be the oldest Chinese text entirely devoted to all the known mathematics that had come before it, a real Chinese Euclid's *Elements*, ending with—if not a proof in the sense of Euclid's axiomatic logic—a serious and appreciated persuasion of the Pythagorean theorem. The Chinese way of doing mathematics was by persuasion through examples and analogies, proof (or, at least, persuasion) no less valid than Euclid's.

Unfortunately, as it is with almost all first millennium BC books, there are no surviving complete copies of the original, which was probably destroyed in a 208 BC book burning ordered by the emperor Qin Shi Huang under a ridiculous pretext of eliminating the obsolete to make way for the new, though it's more likely, as it is with all insecure tyrants, that Qin Shi wanted to erase any evidence that would compare his reign with those of past emperors. We are fortunate, however, to have a 263 AD version of the work that was compiled by Zhang Chang and Geng Shouchang in the first century BC. It has a commentary and supplement by Liu Hui.[6] In the preface, he wrote

> I read the *Nine Chapters* as a boy, and studied it in full detail when I was older. [I] observed the division between the dual natures of Yin and Yang [the positive and negative aspects] which sum up the fundamentals of mathematics.[7]

Lay Yong Lam at the National University of Singapore and Tian Se Ang at Edith Cowan University in Australia, two eminent authorities on Chinese mathematics, tell us in *Fleeting Footsteps* that it is not too wild to consider the possibility that the Hindu-Arabic number system might have come from the Chinese counting rods.[8]

Think of the rods as toothpicks laid out on a flat surface, a counting board. We have several early Chinese mathematical writings that shed some light on Chinese arithmetic—in particular, the fourth or early fifth century *Sun Zi Suan Jing* (*The Mathematical Classic of Master Sun*), which shows numerals as vertical and horizontal rods.[9] Red counting rods were used for positive coefficients and black for negative; they would be placed into (or removed from) counting squares to perform operations of addition, subtraction, multiplication, and division.[10]

It is a place-value system, so, for instance, the number 26,999 would be written as

This is remarkably similar to our Hindu-Arabic decimal system. A problem comes when trying to represent a number such as 2,600,999. There was still no notion of a zero placeholder. The original Hindu-Arabic decimal system had no symbol for zero, but did have a word for empty space (*sunya* in India, and *sifr* in Islam); as with Chinese rods, which also had a word for empty space (*kong*), it too used a blank space as a placeholder.[11] Spaces count, as they did in the Babylonian system, but when writing in a hand script, spacing is often ambiguous. Does the Chinese rod number in figure 3.3 mean 2,600,999 or 260,999?

FIGURE 3.3. 2,600,999 or 260,999?

With this too, the Chinese were inventive. Notice that there is a space (a kind of zero) between the 6 in the ten-thousands rank and the 9 in the hundreds rank; however, just to be sure that the number represents 260,999 and not 2,600,999, the rods are alternating orientation through the ranks. The rods indicating the 6 and the 9 are of the same orientation; that altering skip signifies that there is a zero between the 6 and 9. To write 26,000 poses a slight problem, but the simple solution is to write it as ‖⊥ 一千, which translates to (26 thousand).

Remember how the ancient Babylonian system used a space as a placeholder for zero and two spaces for a double zero? This Chinese system circumvents that need by that clever orientation trick. So, if two zeros were sandwiched between the 6 and 9, the orientations would have indicated so.

How clever! A yin-yang notion of opposites. This effective little trick also works nicely for distinguishing a double zero from a single zero. The number illustrated in figure 3.3 is 260,999. The number 2,600,999 would be represented differently as

Ⅱ⊥　　Ⅲ≡Ⅲ.

There seems to be two spaces between the 6 and 9; that would make it 2,600,999, but to be sure, look at the alternating orientations of the rods.

If there were just one space between the 6 and 9, the orientation of the 2 and 6 would be as it is in figure 3.3. The system is particularly helpful in distinguishing 26,990 from 2,699. The first is represented as

Ⅱ⊥Ⅲ≡,

and the second as

≡Τ≡Ⅲ.

Another important mathematics book is the *Suan shu shu* (*A Book on Numbers and Computations*). Among others, it was discovered in 1983 when archaeologists discovered an ancient tomb in central China that had been sealed since the first century BC.[12] Almost two hundred traditional bamboo strips were found; when connected, those strips would form the *Suan shu shu*, a book, about the same clever counting rods, yet with something more: a matrix system for making arithmetical calculations.[13] The illustrations of figure 3.4 show how 9 would divide 6,538.

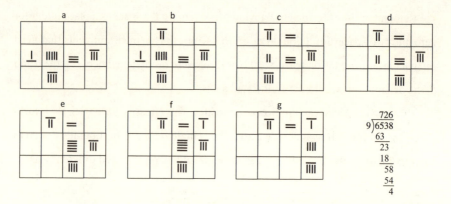

FIGURE 3.4. Dividing 6,538 by 9. Source: Philip D. Straffin Jr., "Liu Hui and the First Golden Age of Chinese Mathematics," *Mathematics Magazine*, vol. 71, no. 3 (1998): 164. Reprinted with permission of the Mathematical Association of America.

In square (a), the first row starts out blank, just as the top line would in our own long division. The second row represents 6,538. The 9 in the third row is placed in the hundreds column under the 5. Notice that, in square (b), the 7 is entered as a result of dividing 65 by 9. And so, in square (c), the 6 and 5 are taken away and replaced by the remainder 2 in the hundreds column. The process continues just as it would in our own long division using our own Hindu-Arabic numerals.

The fundamental operations of arithmetic in this rod system are identical to those of the Hindu-Arabic system. Merchants, scientists, and travelers used the rods in China from the fourth century BC until the abacus replaced the rod system sometime during the sixteenth century. Bags of rods were standard issue to military officers in the seventh century.[14] The *Sun Zi Suan Jing* explicitly details how rod calculations for multiplication, division, and extracting square and cube roots were carried out.[15] Directions for multiplication and division calculations in the *Sun Zi Suan Jing* were the same as those for Hindu-Arabic numbers in al-Khwārizmī's book on arithmetic. The near identical descriptions for computations in the two systems have led some experts to believe that our Hindu-Arabic system may have been transmitted from China to India. So say the two authors in *Fleeting Footsteps*: "It was the only known numerical system which is conceptually identical to the Hindu-Arabic numeral system."[16]

For most ancient cultures, the symbols for the first three numerals are either horizontal or vertical lines, most likely evolved from representations of fingers or sticks. When we reach the symbol for four, we generally do not see four vertical or four horizontal lines, but rather a configuration of lines, possibly four. For some cultures, the transition from parallel line markings to other configurations does not occur before the number six. The Chinese system is one of the oldest, and one in which we can see a logical finger-counting or stick-counting progression.[17] The symbol for 6 should not be six vertical sticks, because it would be difficult to distinguish five vertical sticks from six without counting—the whole point of the numeral symbol is not to have to count. This is very similar to contemporary tallying, where 5 is marked with four vertical lines and a fifth horizontally crossing through.

A child learns the colors of the rainbow long before grasping the meaning of color. Same with number words and the concept of number. Asked to invent a number system never having seen one, you or I might come up with a Greek or Hebrew system. It is simple (almost natural) to invent, yet, like the early desktop computers, awkward to use.

Long before anyone thought of writing numbers using a base system, number writing was done as marks, often batched together in fives. There was no real need for numbers to have individual symbols, as long as the number of batches was not too big. What could "not too big" mean when there were no words or symbols to mark the cardinality of such a "number of batches?" Such a system would be fine for, say ten or twenty marks, but without names or pictures of individual numbers, the system breaks down for large tallies.

Recently, I overheard a conversation between two of my granddaughters: five-year-old Lena asked her ten-year-old cousin Sophie, "Why do I have five fingers on my right hand?" Sophie's answer could not have been better. "So we can count right." Who but a child could come up with such a wonderful cart-before-horse answer?

An Athenian in one of Plato's short dialogues wisely asks how we learned to count:

> How did we learn to count? How, I ask you, have we come to have the notions of one and two, the scheme of the universe endowing us with a native capacity for these notions? There are many other creatures whose native equipment does not so much as extend to the capacity to learn from our Father above how to count. But in our case, God, in the first place, constructed us with this faculty of understanding what is shown us, and then showed us the scene he still continues to show.[18]

Pure mathematics depends on the meaning of "number." Isn't it extraordinary that we understand numbers correctly from almost our first encounters with them, and that we feel comfortable with using numbers before we know what they really are? The Athenian argues

> Recall our very just observation, that if number were banished from mankind, we could never become wise at all. For a creature's soul could surely never attain full virtue if the creature were without rational discourse, and a creature that could not recognize two and three, odd and even, but was utterly unacquainted with number, could give no rational account of things whereof it had sensations and memories only, though there is nothing to keep it out of the rest of virtue, valor, and sobriety.[19]

We can define what we mean by number. But no matter how we define number, its meaning must lead to normal worlds that are consistent with the principles that we have established by design—namely, that what Bertrand Russell, in agreement with my granddaughter, once said should be true, "We want to have ten fingers and two eyes and one nose."

Chapter 4

The Indian Gift

Some low Brahmi numbers (figure 4.1) graphically resemble our low modern numbers. The Brahmi system, however, was conceptually very different. It was not a positional system of powers of ten. Rather, it was closer to an alphabet-based numerical system that requires long concatenated strings to represent even relatively low numbers.

FIGURE 4.1. Brahmi numerals.

At one time, there was speculation that the figures past 4 had come from either the forms of initial letters or syllables of number words of the third century BC Brahmi alphabet. But they may have come from older, untraceable numerical symbols.[1] A more fitting origin is the *Devanagari* script of Sanskrit, initially a spoken language of the Punjab that later branched to the "Vedas" (knowledge), a written medium for religious hymns and invocations generally in verse form or short sentences called *sūtras*. Numerals are integral to those Vedas, which usually referred to achievements of Indian gods who destroyed ninety-nine cities or gave away sixty thousand horses. Some Vedic texts account for sets of numbers as high as a trillion.[2] Later Vedas were considered sacred knowledge, including elaborate astronomy timing accounts for daily sacrifices. And some Vedas used successive powers of ten to describe large numbers.

Unfortunately, due to the harsh subtropical climate, much of pre–first millennium BC Indian mathematics legacy is untraceable. With very few archaeological clues, the origins of Indian numerals must rely on a small wealth of writing that survives almost exclusively in the form of stone inscriptions. So the story of how our numbers evolved is very uncertain. Still, some of those stone epigraphs used decimal place-value numerals, providing some evidence that ancient India was familiar with a kind of place-value numerical system.

It may seem a stretch to say this after examining the Sanskrit words for numbers, but it is possible that some letter combinations of those number words contributed suggestive shapes early in the morphographic history of our current script. (See figure 4.2.)

Hindu-Arabic	1	2	3	4	5	6	7	8	9	0
Sanskrit words	ekah	dvau	tryah	catvarah	pañca	sat	sapta	ashta	nava	suunyá
Sanskrit script	१	२	३	४	५	६	७	८	९	०

FIGURE 4.2 Hindu-Arabic versus Sanskrit.

These numerals give a more complete picture of a place-value system as well as a system that treats zero as a number. In a place-value system, the numerals have different values depending on their position relative to each other. Today's global scientific community has adopted the Hindu-Arabic system. There are, however, minor and major variations in the scripts used in the Middle and Far East. The Eastern Arabic or Indic script is used in present-day Pakistan and Iran. Other systems, such as the Japanese, use both the Hindu-Arabic and kanji characters, the Arabic in horizontal writing and the Chinese in vertical writing. Then, separate from the kanji numerals is a special script called *daiji*, which is used in legal and financial documents to prevent anyone from adding a stroke to turn, say, a two into a three.

Figure 4.3 represents a partial morphography of our modern Hindu-Arabic script, starting with the Brahmi.

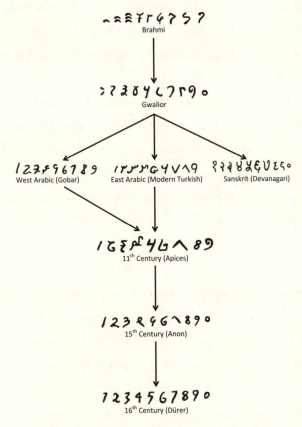

FIGURE 4.3. Modern numerals genealogy. Redrawn from Karl Menninger, *Number Words and Number Symbols: A Cultural History of Numbers*, trans. Paul Broneer (Cambridge, MA: MIT Press, 1969), 418.

History has very few well-defined, unbroken straight lines. The morphography of modern numerals is far more shattered than the flow chart in figure 4.3 suggests. The mystery here is that there is no clearly established, smoothly defined lineage from early scripts to modern ones. The materials and tools for writing, as well as scribing mistakes, must have shaped the numerals beyond any resemblance to their originals.

I suspect, without any verifiable evidence, that finger counting is responsible for the design and evolution of number symbols. Imagine going to market to buy a fish in Ur during Abraham's time. You would probably raise one finger as a signal that you want only one fish, two fingers for two fish. The orientation of your hand could be vertical or horizontal. Thus, the symbol for "2" could be designated by two vertical fingers or two horizontal fingers, which are rapidly sketched as two horizontal lines that over time become morphographically corrupted by faster and faster representations.

We do not know the true origins of the Brahmi system; nor do we know of the many other intermediate paths ignored by the historic record. Did the third century BC Brahmi system come from the Brahmi alphabet, some other alphabet, the old Egyptian numerals, an earlier Indus culture, or from more ancient numerals? Did the Gwalior numerals, a place-value system, come from the Chinese, as the mathematical historian Lay Yong Lam claims?[3] She tells us that by the first century AD the Chinese had a decimal-based place-value system with nine signs, and a concept of zero.

George Gheverghese Joseph tells us in his book *The Crest of the Peacock* that, aside from the Babylonians' clever sexagesimal (base 60) positional system, our modern place-value system is exclusively Indian.[4] And yet Robert Kaplan, in his book *The Nothing That Is: A Natural History of Zero*, tells us that our system was Indian, but originated with the Greeks. Without solid written proof, there is no way of filling in the blanks of history. All we really know is that somehow, in some time and place, the clever place-value idea was transmitted from the Indians to the Arabs and later to the Europeans.

The French mathematician Pierre Simon Laplace confidently claimed:

> It is India that gave us the ingenious method of expressing all numbers by means of ten symbols, each symbol receiving a value of position as well as an absolute value; a profound and important idea which appears so simple to us now that we ignore its true merit. But its very simplicity and the great ease which it has lent to all computations put our arithmetic in the first rank of useful inventions; and we shall appreciate the

grandeur of this achievement the more when we remember that it escaped the genius of Archimedes and Apollonius, two of the greatest men produced by antiquity.[5]

The Eastern Arabic numerals are still used in Arab countries east of Egypt, where they are known as Indian numerals. In Morocco, where Western Arabic (Gobar) is used, the numerals are called Arabic.[6] The apices are similar to the Gobar numerals and seem to be of Indian origin, though there is no direct evidence. The Gobars are known to be Indian, but the apices have always been in question. The early-twentieth-century German historian of mathematics Moritz Cantor claimed that Boethius created the apices from the Gobars and that the earlier Indian numerals found their way to Alexandria before the end of the fourth century, when its commerce connections with India ended.[7] Cantor also claimed that the Indian numerals (without zero) arrived in Christian Europe more than a century before al-Khwārizmī's *Algorism* was translated into Latin; that would have put its arrival in the eleventh century.[8]

Though there are variations from East to West, we must wonder why it is that during their long migration in all directions from India, from culture to culture, country to country, for over fifteen hundred years their basic scripts have remained almost entirely unchanged. The importance here is that though the symbols themselves may look different, each system beyond the Brahmi uses place-values for powers of 10 and a zero. The Brahmi was not a developed place-value system. It had separate symbols for 10, 20, 30, 40,..., 90 and 100, 200, 300, 400,..., 1,000. A Brahmi would not write the number two hundred and twenty-two as ਃਃਃ, which would be done in a place-value system, but rather as Υ☉ਃ. That is because Υ is the Brahmi symbol for 200 and ☉ is the symbol for 20.[9]

The question remains: How did the Western system of numerals with zero come to be? To answer, we should first turn to finger counting, the dust board and the abacus.

At some time in the first half of the second millennium, merchants did their counting and simple arithmetic by finger bending. The merchant would hold up his hands with palms facing outward and indicate numbers according to schemes like the following (figure 4.4):

FIGURE 4.4. Finger counting from a page of Luca Pacioli's *Summa de Arithmetica*, published in 1494.

1. On the left hand;

    To indicate the number 1, half close the 5th finger only;

        ...2, the 4th and 5th fingers only;

...3, the 3rd 4th and 5th fingers only;

...4, the 3rd and 4th fingers only;

...5, the 3rd finger only;

...6, the 4th finger only;

...7, close the 5th finger only;

...8, the 4th and 5th fingers only;

...9, the 3rd 4th and 5th fingers only.

2. Still on the left hand, a different set of symbols would indicate the numbers from 10 to 90. For example, to indicate 10, place the tip of the forefinger at the bottom of the thumb, so the resulting figure resembles the Greek letter δ that, as a number, indicates 10.[10]

Such hand symbols provided nothing more than a numerical gesture language for merchants ignorant of each other's language, for there were no arithmetical calculations to come of it. Hand signals are still in use at the New York Mercantile Exchange, the American Stock Exchange, and other security exchanges, where "open outcry" hand signals indicate buy and sell orders: traders hold up fingers with palms facing toward the body to buy, and fingers facing away from the body to sell in complex gestures that can indicate a wide range of trade possibilities.

I am reminded of an adventure I once had almost half a century ago while traveling to Cabruta on the Orinoco in Venezuela. I woke early one morning on a market day when everyone came to the village square to drink coffee. *Panares*, the indigenous peoples of the region, sold parrots, monkeys, baby ocelots, and river dolphin. I learned that in the Panares language, the words for parts of the body were used to indicate numbers. The word for hand meant five; the words that expressed "other hand" meant six; the words that expressed "both hands" meant ten. And there were other body part expressions such as "foot," and "other foot," and "both feet" that gave higher number values—I don't recall what, but would guess 11, 16, and 20.

The capacity for humans to add and multiply must have begun with some marking scheme, whether from counting fingers, stones, or something more imaginary.

Early stages of counting must have been done concretely, by pointing to the objects one-by-one. Remnants of Aztec languages use numbers such as one stone, two stones, three stones, and so on. There are South Pacific languages that count one fruit, two fruit, and three fruit. In time, however, counting—in terms of such specific groups of objects such as fingers, stones, fruit, and grains—developed to an abstract stage, where the character of the objects being counted was no longer important. This was mathematics. The formation of the idea of number in the abstract sense developed as a result of repeated counting on fingers or by some other marking scheme.

We have strong evidence suggesting that all number systems evolved from counting fingers, toes, and other body parts. Children naturally use their fingers as the set into which they make a one-to-one correspondence with the names of numbers. Perhaps it is essential for arithmetic development.

The Yupno, an Aboriginal tribe living in the remote highlands of New Guinea, count to thirty-three using an elaborate system that counts each finger in a given order, then counts body parts, alternating from one side to the other.[11] There is a definite advantage to the Yupno counting system. When American children count on their fingers they start with a fist, raise each finger in succession, and stop at the final count. In the end, the number of fingers remaining in raised position is the answer. It presupposes no definite ordering: the child could start with any finger and raise any other finger that is not raised, though there are some cultural standards. Since the Yupno system requires counting in a definite order, it has an advantage: the answer is simply associated with the last body part in the count.

Back in the days when writing was inconvenient in the marketplace, finger reckoning was common. Under the historical evidence of the 1920s, the American mathematician David Eugene Smith wrote, "The general purposes of digital notation were to aid in bargaining at the great international fairs with one whose language was not understood, to remember numbers in computing on the abacus, and to perform simple calculations."[12] The only complete record of ancient finger counting in existence is the codex, *De computo vel loquela digitorum* (*On Calculating and Speak-*

ing with Fingers), written by Venerable Bede, an eighth-century Benedictine monk renowned among Medieval scholars for, among other things, his calculation of the varying date of Easter Sunday, which was designed to never fall on the same day as the Jewish Passover. Since all other Church holidays are determined by Easter, Bede's calculations were considered significant. Bede illustrates how one can indicate numbers from 1 to 1 million by simply extending and bending fingers.[13] Finger notation leads to finger counting, which in turn leads to finger computation. Indeed, we don't need to know the multiplication table beyond 5 × 10 in order to multiply two numbers together. Multiplication of small numbers can be reduced to counting fingers, multiplying by 10 and adding 100. For example, to multiply 6 by 8, subtract 5 from both numbers to get 1 and 3. Raise one finger on the left hand and three fingers on the right. Count the raised fingers (1 + 3 = 4) and multiply by 10 to get 40. Now multiply the bent fingers on each hand (4 × 2 = 8) and add the result to 40. You get 48.[14] To multiply two numbers between 11 and 15, subtract 10 from each; represent those two numbers by raising fingers. Count the raised fingers and multiply by 10. Add the result to the product of the number of raised fingers on each hand and then add 100. For example, to multiply 12 by 14, subtract 10 from each to get 2 and 4. Raise two fingers on the left hand and four on the right. Count the number of raised fingers (2 + 4 = 6) and multiply by 10 to get 60. Multiply the number of raised fingers on each hand (2 × 4 = 8). Add 100, 60, and 8 to get 168.[15]

Sixteenth-century texts show how this simple multiplication is carried out when writing is available.[16] They may even suggest the origin of the symbol for multiplication. If you want to multiply 6 and 8, form the compliments, 4 and 2, by subtracting each number from 10. Now write the four numbers on the square grid as pictured. To get the answer, 48, subtract 2 from 6 to get the 4, in the tens column. Then multiply the two numbers in the right column to get the 8.

In *What Counts: How Every Brain Is Hardwired for Math*, Brian Butterworth asks the question, why is the left parietal lobe (the area of the brain where active movement of fingers is concentrated) also the area devoted to calculation?[17] Could it be that moving fingers is as necessary for counting as the eye is for seeing? If so, Butterworth's question—could it be that calculating ability comes from what we do with our fingers?—has an answer. His hypothesis is that it does. To come close to an answer, we need to put together several pieces of the finger puzzle. Wilder Penfield's famous mapping of the motor cortex (the part of the brain that controls motor functioning of the body) showed that the cells that control adjacent body parts are adjacent in the motor cortex.[18]

But there is more. Body parts that require more complex movement take up larger areas of the brain. Smaller body parts that require more complicated movements, such as the fingers, have a larger representation in the motor cortex than larger body parts that require less intricate movements, such as the arms. Another important consideration comes from extraordinarily surprising results of research with people who use Braille (and hence their fingers) to read. They have a larger motor cortex representation in the area serving the finger. Does the same phenomenon happen in the motor cortex of a person who plays the piano? A court stenographer?

So my granddaughter may have been right after all: it may be that we have five fingers on each hand "so we can count right."

The principle of finger counting carries over to pebble markings that, in turn, may have led to sand reckonings and abaci. I say, "may have," only because there is no reliable evidence to support such a claim beyond some late historical legends. But it is a thought to consider. It is far easier to count one hundred pebbles than to count one hundred scattered grazing sheep. And it is far easier to count ten heaps of pebbles, each containing ten pebbles, than to count one hundred pebbles. Egyptian, Greek, and Chinese did their ordinary calculations by such a pebble-counting technique that used different-sized pebbles. Each pebble of one size represented a pile containing pebbles of a smaller size, so that, say, ten pebbles might represent one hundred smaller pebbles. The system evolved into one that had no need to distinguish between sizes, because the pebble counter learned to place pebbles representing tens in a different place from those representing ones.

This may not seem like an exceedingly advanced thought, but it promptly suggests the concept of the abacus. Early abaci were simply pebble-counting schemes, where the pebbles were placed along lines: lines of ones, tens, hundreds,... A pile of four hundred and twenty-three pebbles may not be practical to count, but if each large pebble represented a hundred, each medium-size pebble represented ten and each small pebble represented one, then four large pebbles, two medium-size pebbles, and three small pebbles would represent 423. (See figure 4.5.)

FIGURE 4.5. The lines for pebble-reckoning were drawn in sand. Source: Georges Ifrah, *The Universal History of Computing: From the Abacus to the Quantum Computer* (New York: John Wiley & Sons, 2001), 11.

A dust board, more like a shallow sand box, was used because the methods required the moving of numbers around in the calculation and rubbing some out as the calculation proceeded. As with a grammar student's slate and eraser of a hundred years ago, or on a white board of this century, numbers could be written, moved, or rubbed out in the course of calculation.

Counting boards go back as far as Babylonian times. However, we have no actual specimens, except a few from Greece. In 1846, a white marble tablet (now at the National Museum in Athens) incised with parallel columns was found on the Greek island of Salamis, giving us direct evidence that counting boards using pebbles go back to at least 300 BC. We also have indirect evidence that counting boards go at least as far back as the fourth century BC by virtue of the Darius Vase. (See figure 4.6.)

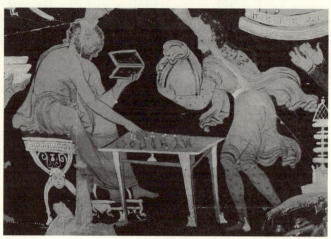

FIGURE 4.6. A royal treasurer of Darius's court and his counting board. Fourth century BC. The man standing is bringing in the booty from the conquered. The counters here are not placed in columns, but directly beneath the symbols for numerical rank. In his left hand is something that looks like an iPad, but it is a pad with a record of the amounts that are found successively on the abacus. Source: Detail from Darius Vase (Museo Nationale, Naples).

For Babylonian, Greek, and Roman counting boards, number representation followed the laws of place-value. There were no symbols for zero; none were needed, for an empty column would indicate that a place rank was held with no numerical value. This idea led to the next: beads threaded on stretched wires—the more modern abacus.

The Roman abacus had metal balls sliding in grooves.[19] The abacus in figure 4.7 has eight decimal positions, marked (from right to left) I, X, C, ∞,... corresponding to units, tens, hundreds, thousands,... all the way up to 1 million. Ignore the first two grooves on the right of figure 4.7 for the moment.[20] Above the markings are single metal balls or counters sliding in their own grooves. For a digit less than 5, the abacist would move the corresponding number of counters toward the letter markings. For a digit representing a number between 5 and 9, the abacist would first move the single counter corresponding to that digit upward to represent 5, and then move other counters on that line upward to complete the representation. All numbers below 10 million could be represented this way. For instance, figure 4.8 illustrates the configuration for the number 5,372.

FIGURE 4.7. A modern replica of a Roman abacus.

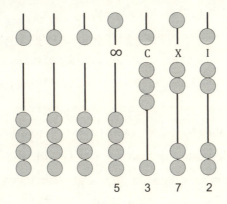

FIGURE 4.8. Configuration for the number 5,372.

Clearly inspired by the abacus, numbers in the tenth-century West were being written as Roman letters of the alphabet in digital order. For example, 5,372 would have been written as V.III.VII.II, mimicking the counters of the four slots of the abacus marked by ∞, C, X, and I in figure 4.7.

Then came Gerbert's counting board. Born in the Auvergne of south-central France in 950, Gerbert was educated in the monastery of St. Gerald of Aurillac. In 967, he left the monastery to travel to Arab-ruled Spain, where for three years he studied mathematics and was exposed to Arabic learning and acquaintance with the Indian numerals. He returned to Rheims, and got a position at the cathedral school to teach mathematics and abacus computation. His career took interesting paths—teaching, then positions: first as abbot, then as archbishop, tutor to the son of Emperor Otto III, advisor to the pope, and ultimately, at the relatively young age of 49, pope. Gerbert d'Aurillac was his real name, but in the ill-omened year before the turn of the new millennium, he became Pope Sylvester II.

The Gerbertian abacus enjoyed a brief period of popularity between the late tenth and mid-twelfth century. Very little is known about the original Gerbertian abaci (none survive), but some recently discovered manuscripts are thought to be exemplifying Gerbert's abacus.[21] The recently discovered *Echternach* manuscript (ca. 1000 AD) in the Benedictine monastery at Echternach in east Luxembourg is one, and the *Computus* manuscript (ca. 1110 AD) written at Thorney Abbey in Cambridgeshire, England, is another.

Looking at a page of the *Computus*, we find that the columns of a counting board were ranked by powers of 10. In figure 4.9, at the top of the sixth column from the right, under $\overline{C}$ (= 100,000), we find a single "counter" labeled 5. (That would be a symbol looking much like $\mathcal{Y}$ on a real medieval counting board.) The original counters of Gerbert's counting board were carved from the tips of horn; hence, possibly because each was in approximately the shape of a cone with an apex, they were called "apices."[22] Those apices were marked with strange symbols that looked very much like our Indian numerals (the fourth tier down in the Hindu-Arabic numerical evolution chart in figure 4.3). Though the carved horns were merely aesthetic designs with no particular arithmetic advantage, Gerbert ardently made hundreds of apices from horn, whereas other abacus artisans after him made theirs out of the materials used in making the counters for the Roman abacus: ivory, metal or glass.

The idea of replacing a group of pebbles of a particular rank with a single object marked by a symbol was novel; those apices were the first to appear in the West.[23] To us, who think in terms of Hindu-Arabic numerals, the Gerbertian counting board may seem to be the natural discovery-consequence of the Roman counting board, but, back then, it was one of those rare grand leaps in the history of symbol progress. Gerbert certainly had heard about the marvelous possibilities of the Arab

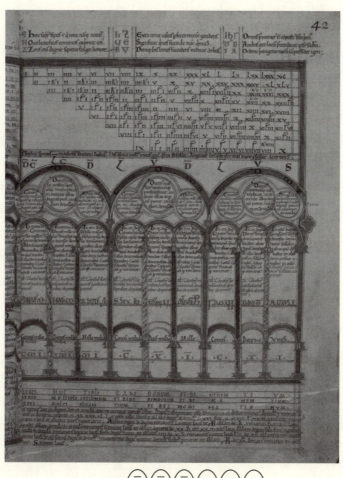

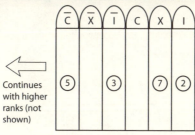

503,072

FIGURE 4.9. Gerbertian counting board showing apices 5, 3, 7, and 2 ranked $\overline{C}(=10^6), \overline{I}(=10^4), X(=10^2)$, and $I(=10^1)$ respectively. Computus manuscript. Written at Thorney Abbey, Cambridgeshire, ca. AD 1110. Oxford, St. John's College MS 17 Fols. 42r. Source: http://digital.library.mcgill.ca/ms-17/folio.php?p=42r&showitem=42r_8Math_1cHinduNumerals#. Reproduced by permission of the President and Fellows of St. John's College, Oxford.

ideas. He must have been aware of the writing computation advantages behind the new system of numerals, but his board was used merely as a computation tool that had no need for written computations. Because the real meaning of the new numerals were not well-understood in the West, Gerbert's disciples went on using those mysterious symbols without regard to their true potential powers.

Those same symbols morphed through a variety of forms, sometimes rotating through different angles, as if it was simply a matter of not caring whether or not it was right-side up or down. This may have been caused by the nature of the counting board, which had no serious orientation, nor any established right way—the abacist may see it differently than someone watching from the other side of the board. In figure 4.9 (the Computus manuscript), the numbers 3 and 8 can be seen as written two different ways on the same manuscript. But whatever the form looked like, the system was the same; with just nine symbols, every number could be represented, and every number could be painted, without taking the brush off the surface.

Between the tenth and twelfth centuries, the abacus board, as either a sheet of parchment or a grooved counting table, with its vertical columns ranking the powers of ten was the main method of studying practical arithmetic in Western Europe. The process of the Gerbertian abacus, called "algorism," was mimicked using quill and parchment.[24]

Chapter 5

Arrival in Europe

1 + 1 is abbreviated into 2, a new and "arbitrary" symbol. "Arbitrary," that is, any other would have done as well. It is 2 that stands for 1 + 1, and not <, 3, ∞, or anything else, because certain Hindoos chose that it should be so.

—Augustus De Morgan[1]

Curiously, few centuries have passed since our wonderful current number system was brought to Europe. There is a dispute over whether or not the person most responsible was Leonardo Pisano Bigollo (ca. 1170–ca. 1250), one of the great mathematicians of his time, whose fame comes mostly from that celebrated problem of how rabbits multiply, a man more memorably known to us as Fibonacci. He was certainly not the discoverer of the answer to the rabbit question, which was asked in ancient India since about the turn of the first millennium to describe the metrical structure and underlying rhythm found in Sanskrit poetry.

A portrait of Fibonacci exists (figure 5.1). It was painted in the middle of the thirteenth century. He looks like a boy with very pleasant features, big eyes, small mouth, and a nose for curiosities. As a young man, Fibonacci traveled with his father around the Mediterranean, meeting priests, scholars, and merchants in Egypt, Syria, Greece, and Provence. He learned the number systems used in trade. Returning to Pisa, he wrote his *Liber abbaci* (*Book of the Calculations*) in 1202 and revised it in 1228.[2] *Liber abbaci* was a book about how to calculate without an abacus, written to convince Western tradesmen that the Arabic numeral system of calculating was superior to the then-used Roman system. It was not the first such Western book to describe Arabic numerals. The *Codex Vigilanus*, the first Western (ca. 976 Spanish) manuscript containing Arabic numerals, had already been available for 250 years (see figure 5.2). And the Latin translation of al-Khwārizmī's *Algorism* had appeared in the twelfth century.

FIGURE 5.1. Fibonacci.

The *Liber abbaci*, however, appeared some 250 years before the invention of printing, at a time when there were no local public libraries, when knowledge spread by word-of-mouth. In Italy, where Hindu-Arabic numerals had an early appearance, it was the *maestri d'abbaco*, practitioners of commercial arithmetic, who peddled the art in private and public tutorials of arithmetic, geometry, and algebra and who wrote mediocre treatises on those subjects as displays of momentous knowledge. They copied from other manuscripts, often changing problem values or tweaking them to disguise their sources. Some seventy-five years after the first appearance of the *Liber abbaci*, students from all over Europe (Bohemians, Poles, Frenchmen, Germans, and Spaniards) visiting intellectual centers such as Venice and Pisa spread the word about Arabic numerals.

Algebra was not generally taught in the universities before the late seventeenth century. In part, that was because universities were primarily places to train for the clergy, or to become a doctor or lawyer. Instead, mathematics thrived in the *bottegas*,

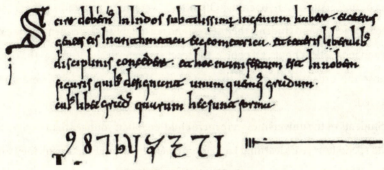

FIGURE 5.2. Codex Vigilanus.

the Northern Italian abacus schools of the fourteenth and fifteenth centuries, where *maestri d'abbaco* taught commercial arithmetic to merchants and artists in the vernacular. The name "abacus school" or "abacus tradition" comes from the fact that students of such schools were studying mathematics in the style of Fibonacci's *Liber abbaci*, not from the abacus as a calculating tool. The term "abbacist" applied to a person skilled in calculating with Hindu-Arabic numerals, as opposed to those who calculated with an abacus.[3] In mid-fourteenth-century Florence alone there were at least 1,200 students attending the roughly twenty *bottegas* of the city. The *maestri* wrote attractively illustrated treatises with the solutions of hundreds of arithmetic and algebra problems that often went beyond the levels taught within the schools.[4]

Charlemagne, the man who conquered almost all of Western Europe, the man who became emperor of the Holy Roman Empire in 800, understood that Europe had lagged behind the Arab countries in science and medicine. He ordered every cathedral and monastery in his kingdom to open schools for public education. Other than geometry and arithmetic, there was hardly any mathematics or science taught at any of those schools. After Charlemagne's death, the curriculums focused on Latin, music, and theology.[5] However, an inexplicable flood of gifted teachers and inquisitive students advanced the medieval curriculum to the liberal arts: first to the *trivium*, comprising grammar, logic, and rhetoric, and later to the *quadrivium*, to include arithmetic, geometry, music, and astronomy. Anyone who successfully went through the *trivium* was a person of great learning.

By the twelfth century, just when the guilds were beginning to form, early versions of universities were opening. Often they were voluntarily formed and organized by teachers and students themselves, and reasonably independent of the cathedral and monastic schools. Of course, the teachers were all connected to the Church because the ordained clergy of the time were the only ones who were already educated.

Students of the universities were mere children, often younger than twelve. They would spend four years learning Latin grammar, for which, if successful, they would be awarded a Master of Grammar degree. A Bachelor of Arts degree requiring successful completion of the *trivium* would be more advanced and more honored. A Master of Arts would require the successful completion of the *quadrivium*, another three years. That was the highest degree possible and very hard to achieve; it was a license to teach, but the pay was poor.

For a long time, Fibonacci's book the *Liber abbaci* was the only known comprehensive source for abacus methods, and so it may seem as though it was responsible for bringing Hindu-Arabic numerals westward.[6] There are current popular books that claim he brought the Hindu-Arabic numerals to Europe. However, since the 1960s several historians have argued that books on calculation involving Hindu-Arabic numerals that were around in Fibonacci's time do not mention the *Liber abbaci*. And, more recently, in 2002, the Danish mathematical historian Jens Høyrup has argued that calculation books spread from Iberia and Provence to Northern Italy, suggesting that Fibonacci was not the person responsible for bringing Hindu-Arabic numbers to Europe.[7]

In the preface to the *Liber abbaci*, Fibonacci writes:

> As my father was a public official away from our homeland in the Bugia customshouse established for the Pisan merchants who frequently gathered there, he had me in my youth brought to him, looking to find for me a useful and comfortable future; there he wanted me to be in the study of mathematics and to be taught for some days. There from a marvelous instruction in the art of nine Indian figures...[8]

Only nine Indian figures are mentioned. So "0" was not included, at least not included as a number.

Høyrup claims that Hindu-Arabic numerals had already been introduced to the Latin culture by the early twelfth century through Iberia and Provence, but that twentieth-century historians credited Fibonacci for the introduction because by his time commercial teaching in Italy was still based on Roman numerals. The confusion seems to rest on the fact that Fibonacci's book consistently uses Arabic numerals in talking about familiar mathematics.

Høyrup goes on to tell us that the Italian abbaco algebra inspiration did not come from Fibonacci, but rather from non-Italian sources and that Italian merchants already had an urgent need for such things as were taught in the abbaco tradition. He continues: "What we can know from the analysis is that the abbaco tradition of the outgoing thirteenth century was *no Fibonacci tradition*, even though it was *already a tradition*."[9]

A hundred years ago, David Eugene Smith and Louis Charles Karpinski, renowned scholars of the history of mathematics, wrote:

> So familiar are we with the numerals that bear the misleading name of Arabic, and so extensive is their use in Europe and the Americas, that it is difficult for us to realize that their general acceptance in the transactions of commerce is a matter of only the last four centuries and that they are unknown to a very large part of the human race to-day.[10]

Smith and Karpinski went on to point out how strange it is that the system had such a struggle to become the standard of the entire world when every other system was so crude and awkward. We tend not to think of our wonderful (isn't everyone's?) system as being so recent, but surprisingly few centuries have passed since the system was passed on to Europe.

In Fibonacci's day, there were numerous texts on calculations. Those texts, however, were scholarly books pitched to people who wanted to learn *arithmetica*, the theory and philosophy of numbers and calculation, or to those who wanted to master the Church calendar.[11] In his PhD research, Warren Van Egmond found that many of those books did not use the word "abacus" in titles, preferring the word "algorisms," which more directly implied an explanation of Hindu-Arabic numerals and their computing algorisms.[12]

Although it is likely that Fibonacci's book had introduced Arabic numerals to some parts of European society, it is also likely that travelers and merchants in Italy already knew those numerals. And surely there were other books written on Arabic numerals as much as half a century before Fibonacci's *Liber abbaci*. The twelfth-century Spanish biblical commentator, scientist, and rabbi Abraham ben Ezra wrote *The Book of the Unit* to describe the Arabic symbols and *The Book of the Number* to describe the decimal system with place-values and zero.[13] His books would not have done much for spreading the word of Arabic mathematics to the public merchants, but they surely helped to get the attention of European scholars. At about the time Ben Ezra was writing his books about place-values and zero, Johannes Hispalensis, one of the main translators at the famous Toledo School of Translators, wrote what is considered to be the earliest known Western descriptions of Indian positional notation in his *Arithmeticae practicae in libro algorithms* (*Book of Algorithms on Practical Arithmetic*).

In his *Short Account of the History of Mathematics*, the early-twentieth-century British mathematician and historian W. W. Rouse Ball tells us:

> Though Leonardo introduced the use of Arabic numerals into commercial affairs, it is probable that a knowledge of them as current in the East was previously not uncommon among travellers and merchants, for the intercourse between Christians and Mohammedans was sufficiently close for each to learn something of the language and common practices of the other. We can also hardly suppose that the Italian merchants were ignorant of the method of keeping accounts used by some of their best customers; and we must recollect, too, that there were numerous Christians who had escaped or been ransomed after serving the Mohammedans as slaves.[14]

At the beginning of the fourteenth century, bankers of Florence were forbidden to use Arabic numerals, and the common use of such numerals did not happen before the sixteenth century. In 1299, the City Council of Florence issued the *Statuto Dell'Arte di Cambio*, which outlawed the use of the Indian system for financial accounts and required all money records to be accounted for in letters, as is required for bank checks today. The motive was security against the fraud of turning a 0 into a 6 or a 9 by simply adding a horn or a tail.

The *Statuto Dell'Arte di Cambio* could not affect the day-to-day dealings in the marketplaces, bazaars, and trading houses where people could make calculations using the Indian system and translating their final tallies into Roman numerals. But it was not so much the Statuto that hindered the new Indian system. Rather, it was the expense of paper and erasable media needed for scribbling calculations. After a calculation, such as long division with steps that had to be crossed out, the paper was useless for the next calculations. The old system required no new expenses beyond acquiring a counting board, abacus, or sand table.

Typus Arithmeticae

In figure 5.3, we find Pythagoras at the counting board and Boethius computing with Indian numerals. Why Pythagoras? Because in the Middle Ages, Pythagoras was falsely considered to be the inventor of the abacus.

Caliph stories provide the backdrops for so many anecdotal yarns that we sometimes forget that they are mostly the Western folk myths of an exotic bygone civilization. Perhaps it is because the Arabs of Baghdad accumulated unbelievable wealth from its conquests and ports on the Persian Gulf, which propelled its trade between China, India, and Russia in the East and all of Europe in the West. The following story of how the Indian numerals came to the Arabs may be apocryphal, as it comes from the *Ta'rikh al-hukama* (*Chronology of the Scholars*), a mid-thirteenth-century book written by Ibn al-Qifti quoting much earlier sources.

The Caliph al-Mansur received an Indian ambassador at the imperial residence in Baghdad.[15] The year was 771 AD. The ambassador's gift to the Caliph was the *Brāhmasphuṭasiddhānta* (*Correctly Established Doctrine of Brahma*), a book written inSanskrit on astronomy by the Indian mathematician and astronomer Brahmagupta that had been written almost 150 years earlier. Al-Mansur was a great proponent of the dissemination of literature and scholarship, and so he commissioned a

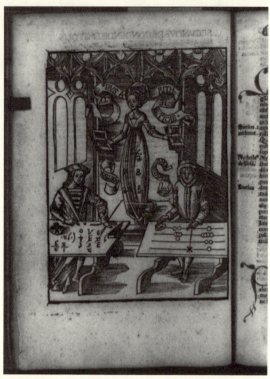

FIGURE 5.3. *Typus Arithmeticae*. A woodcut illustration from *Mararita Philosophica* of Gregor Reisch, which first appeared in 1503 and was used for half a century as an encyclopedia textbook in higher schools. The illustration depicts two calculators (purportedly Pythagoras at the counting board and Boethius computing in the Indian system) in a competition administered by the woman figure personifying Arithmetic. Source: Library of Congress.

translation of the *Brāhmasphuṭasiddhānta* into Arabic.[16] Most likely, the story is legend, because there must have been many sources for Arab astronomy. Legend or not, it was probably this book that prompted Arab scholars to pursue astronomy.

Our zero, as a number as well as a placeholder, probably appeared for the first time in book form sometime close to the year 628 AD. It is in Brahmagupta's *Brahmasphutmasiddhanta* where we first find the rules for using zero with negative numbers ("debts") and positive numbers ("properties"). Brahmagupta marked zero as a solitary black dot to represent the number that results from subtracting a number from itself. Zero was not just a placeholder; for what may have been the first time ever, there was a number to represent nothing.[17]

Not much is known about Brahmagupta. He was probably born in Bhillamala, in southern India, but we do know that sometime in his youth he moved 250 miles southeast to Ujjain to work at the center of mathematical and astronomical research, founded by the sixth-century Indian mathematician-astronomer Aryabhatta. He worked in advanced astronomy, but also developed algorithms for finding square roots and solutions to quadratic equations.

Not much is known about any of the Indian mathematicians of Aryabhatta's time because the historical record of Indian writing of that time is scarce. In those years of ancient India, Hindus believed in a divine or spiritual origin to almost everything, including science. Astronomy and mathematics would have been attributed to Brahma, who created the world, thus bypassing any recognition of the actual humans who were directly responsible for scientific discoveries.[18]

W. W. Rouse Ball felt that once the Arabs left the desert to settle in cities such as Baghdad and Damascus they became subject to diseases for which they had no immunities. At the time, Greek and Jewish medicine was far more advanced than Arabian. For that matter, their knowledge of all science was also far more advanced, based on the works of Aristotle and Galen. So the caliphs encouraged Greek and Jewish doctors to come to teach their science and preserve the traditions of their craft. "The scientific knowledge of the Arabs," Ball said, "was at first derived from the Greek doctors who attended the caliphs at Bagdad."[19]

At around 800, the caliph Harun al-Rachid ordered translations of Greek works into Arabic, and that order was followed by his successor, the caliph al-Mamun, who sent a delegation to the great libraries of Constantinople and India to copy hundreds of Greek and Hindi works. On the delegation's return, an army of Syrian clerks was ordered to work on translating the works of Euclid, Archimedes, Apollonius, and Ptolemy into Arabic and Syrian.

How fortunate that they collected those works, for they are now the only copies extant. Ball also mentions, as a point of curiosity, that Diophantus's works did not seem to be noticed for 150 years after the initial accelerated tradition of translating foreign works into Arabic. By that time, the Arabs were already quite familiar with an algebraic notation process of their own.

Chapter 6

The Arab Gift

Abu Jafar Muhammad ibn Musa al-Khwārizmī's portrait (figure 6.1), popularized by a 1983 Soviet Union postage stamp commemorating the twelve-hundredth anniversary of his birth, shows a bearded man with furrowed brow and dreaming eyes. Isn't it extraordinary that we can know...hmm...what a particular ninth-century person looks like with little knowledge of his biography? The truth is that we hardly know what he really looked like. Al-Khwārizmī, who was the greatest Arab mathematician of his day, learned of the new Indian numbers from the Arabic translation of Brahmagupta's *Brahmasphutasiddhanta*, and wrote a textbook on arithmetic using the new Indian numbers.[1] In around 820 AD, al-Khwārizmī wrote *The Book of Restoration and Equalization*, a book on how to solve equations (in particular, solving for the positive roots of quadratic equations).[2] Its title in Arabic is *Hisab aljabr w'almuqabala*, from which we get the word "algebra." It was translated into Latin sometime during the middle of the twelfth century under the title *Algebra et Almucabala*, and that is how the term "algebra" came to be understood as what it is today.[3] It survives as the earliest Arabic book on algebra.

Al-Khwārizmī's original arithmetic book in Arabic no longer survives. It appeared in Spain at the turn of the twelfth century, and there was translated into Latin by the English Arabist Robert of Chester.[4] That translation (discovered in the nineteenth century) and others of that period were the earliest known introductions of the Hindu-Arabic numerals to Europe, perhaps as much as a century earlier than Fibonacci's *Liber abbaci*.[5]

FIGURE 6.1. Al-Khwārizmī.

Al-Khwārizmī's *On the Calculation with Hindu Numerals*, written sometime near 825 AD, may have been responsible for spreading the Indian system of numeration throughout the Arab world in the ninth century, and then to Europe after a series of Latin translations were made available in the twelfth century.[6]

The years between 786 and 809 in Persia were endowed with scientific and artistic richness. Sometime just before the turn of the ninth century, Harun al-Rashid, the fifth Arab Abbasid Caliph of Iraq, founded a library and translation center in Baghdad that became known as the *Bayt Ul-Hikma* (The House of Wisdom) that turned into a major intellectual center during the Islamic Golden Age of the next five hundred years. Works of astrology, mathematics, agriculture, medicine, and philosophy were translated from Greek, Chinese, and other languages into Arabic at the *Bayt Ul-Hikma*. Al-Khwārizmī worked there, with an interest in all papers originating in India, including Brahmagupta's *Brahmasphutmasiddhanta*, which providentially survived the library worms. While deciphering the mysterious characters and translating the book into Arabic, he discovered something of astounding significance: a way of doing arithmetic that was far simpler than the cumbersome Arab technique.

Until then, Arabs across Mesopotamia had been doing their arithmetic by using either finger counting, or the abacus, or the complicated Roman numeral system, or the messy scheme of writing numbers as words, even when making elaborate calculations of star positions.

In the Hindi writings of the *Brahmasphutmasiddhanta*, al-Khwārizmī might have seen a superb way to easily represent any counting number that could ever be needed with just ten symbols. He might have known about the Babylonian sexagesimal (base 60) system, and thereby seen a way of representing a decimal system. Although decimals were not in vogue in that Early Middle Ages Arab world of commerce, he might have seen the idea as a brilliant vanguard system that deserved, if not the world's attention, then at least serious scholarly interest.

He might have seen that strange black dot that meant nothing, the quantity of nothing. Anyone reading the *Brahmasphutmasiddhanta* would have been baffled by it. Reading on, he might have been inspired by the concept of having numbers that mark negative values as debt. A whole new infinite collection of objects entered in the world of thought, objects symbolizing quantities less than nothing—the negative numbers.

There is a dubious story that al-Khwārizmī traveled to India, where he came in contact with Brahmagupta's mathematical manuscripts. A more likely story, however, is one about the Indian astronomer Kanka, who visited the House of Wisdom in Baghdad in 770 AD, and brought with him many manuscripts from India, including the *Brahmasphutasiddhanta*. This makes some sense, as al-Khwārizmī was a scholar in the House of Wisdom who wrote his book *On the Calculation with Hindu Numerals* some fifty years after Kanka's visit. That book was largely responsible for spreading the Indian system of numeration throughout the Arab world as well as Europe.

At that time, the Arabs had no numeral system of their own. In geographical regions of the Arab world where both Arabic and Greek were spoken, the only system used was either the Greek's alphabetical or one derived from the Greek model using mostly written Arabic words for numbers. The new numerals were at times described as Indian and at other times Arabic. Fibonacci clearly called the nine figures Indian in first lines of the first chapter of his *Liber abbaci*, which translates as:[7]

The nine Indian figures are:

9 8 7 6 5 4 3 2 1.

With these nine figures, and with the sign 0 which the Arabs call zephyr any number whatsoever is written, as demonstrated below. A number is a sum of units, and through the addition of them the numbers increase by steps without end. First, one composes from units those numbers which are from one to ten. Second, from the tens are made those numbers which are from ten up to one hundred. Third, from the hundreds are made those numbers which are from one hundred up to one thousand. Fourth, from the thousands are made those numbers from one thousand up to ten thousand, and thus by an unending sequence of steps, any number whatsoever is constructed by the joining of the preceding numbers.

The confusion was quite possibly caused by the great variation of scripts used to represent the nine figures. By the next century, however, many of the scripts had converged to a standard very close to the one we use today.[8] Yet Arab astronomical tables continued to use alphabetic numerals for centuries after the Arab conquest. There were no consistent uses of the Hindu-Arabic numerals in Islamic numerical jottings.[9]

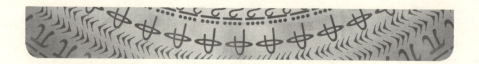

Chapter 7

Liber Abbaci

In the ninth century, the Indian figures were still too new and too weird to spread far from the monasteries and scholarly hubbubs. After all, Europe did not know of a numeral system with a *zero*, that single symbol that could be used to write an infinite range of numbers and at the same time represent nothing. The Babylonian system didn't have one; neither did the Greek, nor the Roman.

It was not as if there were no commerce and travelers to bring the numerals to Europe. There were plenty. It was that beast, zero—the stranger that caused enough suspicion to slow universal acceptance of the new system for more than three hundred years. Today, we accept innovations with such great speed that we hardly notice how they radically affect our lives—the computer chip, cell phone, GPS, movies on demand, medical instruments that extend life. It is staggeringly unbelievable that it took more than three hundred years for Europe to catch on to one of the greatest ideas ever devised to simplify human life. Three hundred years! Where were the Galileos, the Decarteses, or the Newtons of the High Middle Ages?

The difficulty is in distinguishing placeholder and number. Accepting zero as a number representing the absence of quantity would have been a fantastically daring idea. The number two, for example, is fairly easy to comprehend. It represents "two-ness," or the number that represents the counting of two objects. But "zero-ness"? A number that counts no object? What could that possibly mean? The idea of zero as a placeholder, however, is intricately connected to the notion of zero as a number that represents nothing. The confusion is that the symbol for indicating when a position is empty is the same for the number that indicates a count of *no objects*. The Indian idea of nine symbols would not work without a tenth symbol to indicate when a position had no entries. That was part of the problem with the Babylonian place-value system.

Fibonacci, however, was addressing the merchants at the docks, and in the marketplace as well as the court. In his *Liber abbaci*, he implies that the Indian numerals were new to those merchants, for he wrote that he found out about the system as a boy when his father (a public notary) took him to Bejaia (in present-day Algeria), where he learned the art of abbaco.[1] He wrote that the "Latin race" was lacking knowledge of the Indian method of arithmetic, and that other common arts of reckoning, such as the algorism and the apices, are "a kind of error in comparison to the method of the Indians."[2] These were his honest beliefs.

It might seem surprising that Fibonacci, one of the greatest mathematicians of his time, did not show any knowledge of earlier works that clearly speak of the Indian art of reckoning with the nine figures.[3] Shouldn't he have known about al-Khwārizmī's *On the Calculation of the Indians*, which had been translated into Latin by the early part of the previous century? Did he not know of the 976 AD manuscript at the Monastery of St. Martin at Albelda in Spain that said, "We must know that the Indians have a most subtle talent and all other races yield to them in arithmetic and geometry...?" Did he not know about the manuscripts translated by the Toledo School in northern Italy? Did he not know about the Gerbertian abacus that was based on the nine Indian figures? Wouldn't he have known about the Greek-Latin translation of Euclid's *Elements* written somewhere in Tuscany that used the Eastern forms of the Indian numeral script? Notaries just a hundred miles from Fibonacci's hometown of Pisa were already using Indian numerals.[4] From Toledo to Lyon to Munich to Ireland, Latin books on calculation mentioned the numeration of the Indians with nine letters and of how it represents all the numbers.

However, the question of who introduced and influenced the practice of reckoning with Indian numerals to Europe has no simple answer. The evidences are numerous and diverse. Fibonacci was educated as a merchant in Pisa. In school, he was taught to write and count with Roman numerals on an abacus board. In his apprenticeship, he learned how to calculate the prices of goods, how to deal with weights and measures, and how to convert equivalent values of money. By the time he arrived in Bejaia, he was already able to deal with the customary commercial arithmetic of the abacus board. After learning the methods of Indian numerals and their related arithmetical operations, he recognized the advantage of Indian reckoning over those used in Pisa. On returning to Pisa, he had no reason to study Latin texts on the Indian system he had already learned in Bejaia. This may explain why Fibonacci did not mention any knowledge of earlier works that clearly speak of the Indian art of reckoning with the nine figures.[5]

Until recently, Fibonacci's *Liber abbaci* was incontestably recognized by medievalists to be the inspiration for introducing modern arithmetic to the West. In 2004, the mathematics historian Raffaella Franci credited the *Liber abbaci* as "the most important source for abacus teaching in Italy."[6] Two years earlier, another eminent historian, Elisabetta Ulivi, claimed that abacus teaching texts, written in the Tuscan vernacular, were taken from the two sources attributed to Fibonacci, the *Liber abbaci* and the *Practica geometria*.[7] And back in 1980, Warren Van Egmond catalogued a great many abacus texts written up to the middle of the fourteenth century that were directly descended from Liber abbaci, lending evidence to the spread of Indian numerals in Italy and to their links with Fibonacci.

Then (sometime before 1989) came Gino Arrighi's discovery: a book he found in the Biblioteca Reccardiana in Florence, *Livero de l'abbecho* (*Book of the Abacus*), written in the Umbrian vernacular.[8] There is no doubt that the *Livero* was written in 1289 (plus or minus a year) in Umbria. It is an anonymous text: the earliest extant abacus text in the vernacular, a book that might have been modeled on an earlier version, introduces itself by the words

> *Questo ène lo livero di l'abbecho secondo la oppenione de maiestro Leonardo de la chasa degli figliuole Bonaçie da Pisa.*
>
> This is an abacus book that seconds the opinion of master Leonardo from the house of sons of Bonacci of Pisa.[9]

From this, and other credible evidence, Franci advanced the idea that whoever the *Maestro Umbro* author was, he might have altered the writing to fit the needs of his readers. Fibonacci might be the *Maestro Umbro* himself, and perhaps the book is really the lost *Liber minoris guise* (*Book in a Smaller Sense*), a book we know Fibonacci wrote because he referred to it in his *Liber abbaci*. If true, the idea that he is the founding father of Western arithmetic would be clinched. But Jens Høyrup argues that if we alertly read past the introduction, we should "discover that it contains material that is definitely not from Fibonacci."[10] Franci argues, just because the first part of *Livero de l'abbecho* is not from the *Liber abbaci*, that does not mean that it does not bear a resemblance to the *Liber abbaci*.[11]

Respectful of Høyrup's careful reading, Franci altered some of her original views on the nature of Fibonacci's contributions to assert that abacus "authors may have had access to Arabic sources different from those used by Leonardo."[12] She is currently studying two abacus treatises written in Pisa at the end of the thirteenth century or the beginning of the fourteenth century that are closely inspired by the first eight chapters of *Liber abbaci*.[13]

There is no doubt that the introduction of Indian numerals to the West took place from the late tenth century onward; that does not automatically imply that Hindu ways of calculation were introduced before Fibonacci. But, then again, according to another eminent mathematics historian, Charles Burnett, there were many abacus texts of the twelfth century to indicate that Fibonacci was not a pioneer.[14]

Fibonacci told us in the prologue of his *Liber abbaci* that he learned the nine Indian figures used in trade when traveling with his father, meeting merchants in Egypt, Syria, Greece, and Provence. Provence? Wasn't Provence in Western Europe? So how could it be that trade with Provence did not inspire Italian abacus arithmetic?

Høyrup blames what he calls the "principle of the great book," which declares that every book either contains its own originalities or owes its opinions to some famous book that no longer exists. He writes:

> Certain passages in the *Liber abbaci* show that the beginnings of abbacus mathematics must be traced to an environment that already ex-

isted in Fibonacci's days—an environment he knew about and of which he can be regarded an extraordinary early exponent, but no founding father.[15]

Fibonacci did not contribute anything to Indian numerals that was not already known. He was, however, an excellent expositor of new and difficult concepts. Perhaps his talent as an expositor influenced a diffusional migration of the new system from Italy into the rest of Europe. By the middle of the thirteenth century, there were several Latin texts introducing the new system to northern Europe. For example, the *Carmen de Algorismo* was a very popular treatise, written by the French Minorite friar Alexander de Villa Dei in 1240 that explained the methods of computation in 244 verses of dactylic hexameter:

> Here begins the algorismus.
> This new art is called the algorismus, in which
> Out of these twice five figures
> 0 9 8 7 6 5 4 3 2 1
> of the Indians we derive such benefit.[16]

The Indian numerals were popular among the learned in the twelfth and thirteenth centuries because they frequently appeared in monastic manuscripts. A Latin translation by John of Seville, a member of the Toledo School of Translators, appeared soon after Robert of Chester's, and then, in 1143 an abridged version of Robert of Chester's book was catalogued at the library of the Salem Abbey in southern Germany, the oldest evidence that al-Khwārizmī's *Algebra* made its way to northern Europe.[17] There is also Johannes de Sacrobosco, who taught at the new University in Paris and in 1240 wrote *Algorismus*, a textbook on the Indian numerals and how to calculate with them, which was widely used all over Europe. (See figure 7.1.)

So it seems as if the news of the new numbers migrated all over Europe for two or three centuries before Fibonacci wrote his *Liber abbaci*. News, but not practice, and therefore, not much use. One possible reason is that it was misunderstood. There were attempts at adapting the Roman numerals to a place-value system. Roman script was often used in a place-value system without regard to the notion of a zero.

The number 16 in the Roman system, for instance, would be XVI; this uses place-value in the sense that XVI is not the same as XIV. Remember that the Romans also had a counting board that distinguished the tens column from the fives from the units.

> *Que funt tales .0.9.8.7.6.5.4.3.2.1. Decima uero dicitur teca, uel circulus uel cifra uel figura nihili quia nihil fignificat, ipfa tń locú tenés dat aliis fignificare ná fine cifra uel cifris purus non poteft fcribi articulus.*

FIGURE 7.1. A passage from a 1523 copy of Johannes de Sacrobosco's *Algorismus*. In translation from the eighth line down, we have: *Know that corresponding to the 9 units there are 9 number symbols, as follows: 0 9 8 7 6 5 4 3 2 1. The tenth is called theca or circulus or cifra or figura nihili, because it stands for nothing. But when placed in the proper position it gives value to the others.* Source: The Tomash Library on the History of Computing.

How is it that medieval Europe failed to consider the value of an Indian-like system, when merchants and accountants should have seen evidence of it in the form of counting boards, and place-values in abacuses, and spoke of numbers in a conceptually place-value pattern every day in the market place?

What was it that kept them from recognizing the benefits of the Indian system? One possible answer is that it may have been more daunting than we think. Imagine how hard it might be for us to learn to use, say, a Hebrew numbering system after being so comfortable with the Indian system. You know the benefits of a new system only after you have had a chance to use it.

By the end of the eleventh century, news of the Indian system was all over Europe in the form of counters of the Gerbertian abacus marked with Indian numerals. So why do we give so much credit to Fibonacci for introducing the Indian numerals to Europe?

The debate over the origins of modern numerals had been whirling for almost a hundred years before Smith and Karpinski published *The Hindu-Arabic Numerals* in 1911. Smith and Karpinski say that the "general acceptance [of Indian numerals] in the transactions of commerce is a matter of only the last four centuries."[18] Webster's Dictionary lists 0 through 9 as Arabic numerals, yet we have a fragment of a 662 AD manuscript by Bishop Severus Sebokht of Nisibus—the Syrian scholar who lived in the convent of Kenneshre on the Euphrates—that suggests a Hindu origin. That fragment, which is now in the Bibliothèque National de France (MS Syriac [BNF], No. 346), is the earliest-known extant reference to Hindu numerals outside of India:

> I will omit all discussion of the science of the Indians,... of their sub-
> tle discoveries in astronomy, discoveries that are more ingenious than
> those of the Greeks and the Babylonians, and of their valuable meth-
> ods of calculation which surpass description. I wish only to say that this
> computation is done by means of nine signs. If those who believe, be-
> cause they speak Greek, that they have arrived at the limits of science,
> would read the Indian texts, they would be convinced, even if a little
> late in the day, that there are others who know something as well as
> they.[19]

So the nine signs surely were known to Severus Sebokht, one of the leading popular transmitters of seventh-century Greek philosophy and science in Syria. The
translated fragment above, besides its wry comment about others who speak Greek,
claims that we owe the idea of expressing all numbers by nine signs to the Hindus.[20]

Sebokht assumed that the system came through Persia on its way west from
India.[21] More recently (if we can say that 1977 is recent), the historian of medieval
astrology Richard Lemay wrote that al-Khwārizmī's *Arithmetic* was translated into
Latin in three different versions during the twelfth century, along with his *aljabr
wa'l-muqabala*. "Al- Khwārizmī's *Astronomical Tables*," he writes, "was the most notable single channel through which the Hindu-Arabic system of numerals was made
known to the West," most likely from al-Khwārizmī's *Arithmetic*, where the Indians
are credited for the originality.[22]

Among the Arabs, the nine numerals were also called "Indian letters" or "figures" (*al-huruf al-hindi*). One of the few positively dependable sources is the tenth-
century account in *Meadows of Gold and Mines of Gems*, the thirty-volume life's
work of the Arab adventurer and storyteller Mas'ūdì (Abu'l-Hasan 'Ali) published
in 957 AD.[23] Mas'ūdì wrote in the first chapter that he chose the title of his book
"in order to excite a desire and curiosity after its contents, and to make the mind

eager to become acquainted with history."[24] Mas'údì was a curious and inquiring fellow who collected histories of the Persians, Hindus, Jews, Romans, and the cultures of Eastern civilizations. Born in Baghdad, Mas'údì traveled to India, Ceylon (present-day Sri Lanka), across the Indian Ocean to Madagascar, and up the Red Sea back to Egypt, Palestine, and Syria. In 926, we would have found him near the Sea of Galilee in Tiberias; by 943, near the Mediterranean in Antioch or in Cilicia; and two years later, in Damascus.[25] He dedicated his book as "a present to kings and men of learning. Having treated in it on every subject which may be useful or curious to learn, and on any knowledge which arose in the lapse of time." Mas'údì used the Hindu-Arabic numerals throughout his work, starting with a description of Hosaïn, an astronomer who compiled astronomical tables and related wildly inaccurate facts about the circumference and diameter of the earth.[26]

Albelda de Iregua is a small town in northern Spain where the San Martín de Albelda Benedictine monastery now stands in ruins. In its tenth-century heyday, it was the most important and advanced cultural center of Western Europe, quite possibly because it was directly on the trade routes along the Ebro, connecting Castile in the northwest with the Mediterranean. It had a copious library of the richest collection of medieval Spanish literature available to the West, including the first record of Arabic numerals from 1 to 9 in Western Europe. The *Etymologiae*, a Latin manuscript written in the monastery in 976 by Isidore of Seville, already showed somewhat modern forms of the numerals, except for 4. Forms evolved over the years, while basic styles and distinctive features converged toward a standard. Pinpointing an exact time of convergence may be impossible, though my hunch is that passing mathematician refugees were responsible.

The earliest extant Arabic work on Indian arithmetic is the *Kitab al-fusul fi'l-hisab al-hindi* of Abu'l-Hasan Ahmad ibn Ibrahim al-Uqlidisi, composed in Damascus in ca. 952 AD. The earliest Arabic examples of the use of Indian numerals are two legal documents written on papyrus in Crocodilopolis, a city (the oldest city in Egypt), named by the Greek explorers who witnessed its inhabitants worshipping a live crocodile. Those documents mark dates 873–874 and 888–889 AD in Arabic numerals; the next oldest examples are not earlier than the eleventh century. By the twelfth century, there was a distinct difference between the writings of Hindu-Arabic numerals in the Western and Eastern parts of the Islamic world, as described by the Moroccan mathematician, ibn al-Yasamin, who died ca. 1204. The earliest known book using the Hindu-Arabic numerals written in vernacular Italian, *Libro di nuovi conti* (*The Book of New Calculations*), had been written around 1260, but it no longer survives.[27]

With all this documentation and evidence, what can we conclude about the origins of our numerals? Were they Indian? Arabic? Chinese? French? The origin of the Hindu-Arabic numerals has been argued by experts for nearly two centuries. One such expert was the French mathematician and historian Michel Chasles, who patriotically argued an absurd case for a French origin based on obviously fake documents.[28]

Chapter 8

Refuting Origins

Suffering a need for documents he could not validly collect, Denis Vrain-Lucas resorted to stealing antique paper from several libraries in Paris by cutting the endpapers of old books. Using special self-made inks, he carefully imitated diverse handwritings, and sold forgeries (letters and documents) to unsuspecting manuscript collectors.

He was a law clerk and amateur historian with a genuine passion for collecting manuscripts of great historical importance. Over a sixteen-year period starting in about 1855, Vrain-Lucas sold over 27,000 autographed forgeries, many to his favored mark, Michel Chasles, who paid hundreds of thousands of francs over a nine-year period beginning in 1861. Letters autographed Pascal, Galileo, Descartes, Newton, Rabelais, and Louis XIV might have been believably authentic, but Vrain-Lucas had developed such a respectable prominence in the manuscript collection world that he was able to pawn off the ridiculous as well.

The naïve Chasles bought Cleopatra's signed letters (in French!) to Mark Anthony, a signed letter from Alexander the Great (also in French!), and other letters between Pascal, Newton, and Galileo all in French proving that Blaise Pascal discovered the law of universal gravitation. Newton's description of universal gravitation in the *Principia*, was published twenty-five years after Pascal's death, so any such letter signed by Pascal would have been, indeed, astonishing. And yet, in 1867, Chasles stood before the French Academy of Science to present evidence in consequence of his treasured letters while some members of the Academy wondered in disbelief, and others, enwrapped in national pride, assumed it true. In 1869, Vrain-Lucas stood trial for forgery, and was sentenced to two years in prison, but with no pressure of restitution to Chasles.[1] Yet, even after convincing evidence showed that the manuscripts were fraudulent, Chasles persisted they were genuine.[2]

One member of the Academy had unwavering doubts. His full name was Count Guglielmo Libri Carucci dalla Sommaja. For most of the 1840s, Chasles and Libri argued fiercely with each other, mostly on the national origins of numbers and the origins of algebra, at meetings of the French Academy of Science.[3]

Chasles argued that by the fifth century, France already had a decimal place-value system for computations documented in Boethius's *Arithmetic*, which seemed to use a multiplication table with Arabic numbers. Later scholarship clearly doubted that the original text used Arabic numbers. But Chasles argued that Fibonacci's *Liber abbaci* was influenced by Arabic authors.

Meanwhile, Chasles's adversary, Count Libri, had just published a volume of his *Histoire des sciences en Italie*, which addressed the question of Indian origins of arithmetic and positional notation used by Arab authors.[4] Chasles challenged Libri's view that their modern number system came to Europe by the work by the Italian Fibonacci, arguing that it came to Europe by the Frenchman François Viète (1540–1603).[5] It was a public feud, fought with strongly sociopolitical antagonisms.

At 38, Libri was appointed Chief Inspector of French Libraries, which stirred in him an old joy of handling rare books as well as an uncontrollable urge to steal rare manuscripts by the cartload. By the age of 45, a warrant was issued for his arrest. He fled to London with over 20,000 rare books and manuscripts, among which were some of the books he stole as a young man from the Biblioteca Medicea Laurenziana in Florence. It may be hard to understand such odd book thieving among prominent people of academic responsibility, but such behavior was not so abnormal in nineteenth-century France.

For much of the nineteenth century, the Indian origin of positional decimal notation had been challenged.[6] Then, in 1907, an English amateur indologist official working in the Department of Education of the Government of India in Shimla, the summer capital of the British Raj, published an article in the *Journal of the Asiatic Society of Bengal*.[7] In that article, George Rusby Kaye (1866–1929) claimed that the numerals and the decimal place-value system could not have been Indian in origin. His argument, in part, comes from a misinterpretation and dating of the Bakhshâlî Manuscript, a birch-bark document in Sanskrit and Prakrit unearthed by a farmer in 1881 near the village of Bakhshâlî (now in Pakistan). The manuscript was found in fragments, just seventy leaves of birch bark from what may have been many hundreds that have deteriorated through careless handling.[8]

A variation of our numerals, as well as a place-value system, is unarguably in the Bakhshâlî Manuscript. The date of the manuscript, however, had been in dispute. Some scholars put it at 400 AD, others at 700 AD. Kaye claimed 1200 AD was the likely date, but in his highly influential 1907 article wrote, "We can go further and state with perfect truth that, in the whole range of Hindu mathematics, there is not the slightest indication of the use of any idea of place-value before the tenth century AD."[9] He implied that the notation in question was Arabic in origin. Either he did not understand the positional notation, or, possibly, he had the British Raj's colonial interest to keep the Indian origin unlikely.[10] Recent scholarship estimates the date of the Bakhshâlî Manuscript to be between 200 BC and 300 AD, based on the belief that its language had been extinct since the third century.[11]

Agathe Keller, at the Université Paris VII-CNRS, gives this assessment of Kaye:

> We are here in one of these strange but familiar moments that history
> of science encounters, usually in stories of science: the denial of facts.
> How can we understand G. R. Kaye's attitude? He certainly had access
> to texts that discredit such a claim.[12]

Aryabhatta (476 AD) knew of the decimal system, and so did Brahmagupta (550–606 AD). The *Vyasa-Bhasya* is a fifth-century or sixth-century discussion on yoga written in Sanskrit by Vyasa. In it, there is an illustration by mathematical analogy: "the same figure '1' stands for a hundred in the place of a hundred, for ten in the place of ten, and for a unit in the place of unit." Therefore, the Hindus knew of the decimal system long before the Arabs.[13] It was known in China. So how could Kaye have denied the Indian origins of our number system?

In subsequent papers, Kaye wrote that the history of Hindu-Arabic number representation was complicated by the existence of so many forgeries of the time.[14] He suggested that the West knew much of the mathematics formerly attributed to early Indian mathematicians. Keller believes that by "Western knowledge," Kaye meant Greco-Latin wisdom conveyed to the West by Arabic scholars. It seems that Kaye was tormented by the idea that place-value originated in early Sanskrit texts.

All this would have been fine had Kaye's theses not been taken so seriously. His articles were popular among indologists, and quoted by very respectable mathematics historians, quoted even by the highly distinguished early-twentieth-century scholars D. E. Smith, L. C. Karpinski, Florian Cajori, and George Sarton. As late as 1927, the eminent Indian mathematics historian Bibhutibhusan Datta wrote:

> The significance and importance of a publication like the present one are apparent to all lovers of the history of science. And their thanks are certainly due to Mr. Kaye for the great amount of pains that he has taken in explaining and editing the Bakhshâlî manuscript.[15]

It seems clear that the ten numerals used, including zero for an empty position, came to us from the Indians by way of the Arabs. So from here on in this book we shall call them Indian numerals.

Al-Khwārizmī described the Indian numerals as Sanskrit symbols, but his treatise, *On the Calculation of the Indians*, was not translated into Latin and not brought to Europe before the thirteenth century, when merchants still did their daily reckoning with Roman numerals. This may have been the cause of the confusion in the origin of our numerals. Indians certainly did visit lands further west of Syria; Brahmins visited Alexandria as guests of the Roman court in 470 AD. Yet, at such an early time, the numerals were neither viewed as intellectual nor scientific jewels, "but rather like the numerals of alien peoples that become known in the harbours and ports."[16]

Whatever the truth, it is quite likely that sometime in the fifth century, Indian numerals had come to Alexandria via a trade route through Syria. From Alexandria, which had significant connections to Europe, the numerals moved westward.

In the beginning, whenever that was, the nine numerals took on a miscellany of forms; however, by the beginning of the thirteenth century when Fibonacci wrote his *Liber abbaci*, those forms, with a few exceptions, were beginning to settle down to the shapes we see today.[17] Europeans, who at that time did their laborious numeration and calculations using Roman numerals, were given a gift: the awareness that just ten symbols were enough to represent any number in the conceptually infinite collection of all numbers. The Roman numeral system could not do such a thing, for it would require a new symbol for each power of ten.[18]

Of course, no matter what system they used, people with time and tenacity could always calculate. They always had! Long before the magnificent system of nine numerals with its place-values, zero, and easy arithmetic, the abacus had been giving merchants, astronomers, and mathematicians in the East an easy tool for making hard calculations. In one form or another, for almost five thousand years, it had been used as a reasonably efficient calculating tool. It spread westward in the tenth century.[19]

Arithmetic first came from the marketplace and was later elevated to tackle astronomy. Numbers, the language of merchants, must have come from the words used to depict numbers: one, ten, a hundred … These were mere words before they became symbolized by any kind of notation. An easy way of describing large numbers came with the invention of zero; then we could say one-zero, two-zero, one-zero-zero, and so on. A single number can be used and reused to represent infinitely different numbers. With it, we get the idea for potentially writing infinitely many numbers by strings of numbers that we already have.

Though word of Indian numerals spread widely through the merchant and trade population, they were snail-slow in becoming the standard. "Not until the sixteenth century had the new numerals won a complete victory in schools and trade. Even as late as Nikolaus Copernicus' famous work, *De revolutionibus*, published in 1543, the year of his death, one finds a strange mixture of Roman and Indian numerals and even numbers written out fully in words."[20]

There have been many scrupulous studies on the origins of our system, but even after a hundred years of scholarly wide-reaching research, we are left with only sketchy guesses of its beginnings and evolution (figure 8.1).

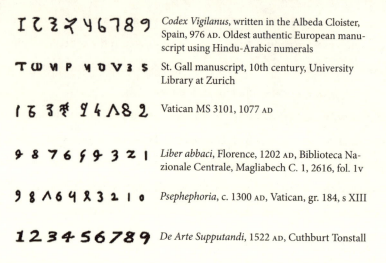

Codex Vigilanus, written in the Albeda Cloister, Spain, 976 AD. Oldest authentic European manuscript using Hindu-Arabic numerals

St. Gall manuscript, 10th century, University Library at Zurich

Vatican MS 3101, 1077 AD

Liber abbaci, Florence, 1202 AD, Biblioteca Nazionale Centrale, Magliabech C. 1, 2616, fol. 1v

Psephephoria, c. 1300 AD, Vatican, gr. 184, s XIII

De Arte Supputandi, 1522 AD, Cuthburt Tonstall

FIGURE 8.1. Morphography of Indian numerals. For a more comprehensive account of the variations and morphographic changes of numerals in the East and West, see Charles Burnett, "Indian Numerals in the Meditarranean Basin in the Twelfth Century, with Special Reference to the 'Eastern Forms,' " *China to Paris: 2000 Years' Transmission of Mathematical Ideas*, ed. Yvonne Dold-Samplonius, Joseph Dauben, Menso Folkerts, and Benno Van Dalen (Wiesbaden: Franz Steiner Verlag, 2002), 237–284.

We can speculate on the morphographics of our numerals. As with almost every culture, the writing of low-value numerals begin as either dots or lines, very likely the result of what instruments were at hand at the beginning—knife, chisel, twig, or reed. Writing on wood, stone, or clay would have been slow. When writing in ink on papyrus, parchment, or paper, however, speed is gained by not having to lift the brush or pen. Every one of our modern numbers from 1 to 9 is, as if by design, a single mark of the pen. The natural morphography that produced our "2" seems to be $=$, 2, 2. There is no established evidence for this, but speculation that this is true seems right. The diagonal line appears to be an unintended dragging of ink from top bar to bottom so as not to waste time in lifting the pen or brush high enough off the parchment. Likewise, "3" probably came naturally from the quick scribing of three horizontal bars. At times these bars were written vertically, in which case the numbers would look very much like our own, only rotated by 90 degrees.

But "4"? Where did that come from? At first glance there are only three strokes—the vertical, horizontal, and diagonal. Looking at it as two angles ˧r—that is, four short lines ˗ˡ˗ quickly scribed—it winds up as a single unbroken mark ✦. The diagonal comes from the unintentional dragging from the vertical mark to the horizontal.

Oddly, the number of strokes in our modern numeral script has no direct connection with the cardinal number itself. Early morpha of number symbols are no longer traceable. For example (see figure 8.1), from the tenth through sixteenth centuries, the number 5 looked very much like an "h" in different orientations, sometimes upside down, and at other times not. Before the sixteenth century, the number 4 had no resemblance to our modern 4.[21]

History moves by unintended consequences, often by chance and coincidences that are hard to foresee and tough to control. Sunlight warms the slime of shallow pools of water to slowly create the biochemical conditions to start life on earth. Earthquakes bury civilizations. Nations change by unintended drivers that can neither foretell the astuteness from the foibles of their leaders nor the rewards from the pitfalls of their decisions. World War II may not have happened had the peace treaty of World War I been more understanding of its consequences. Or, had Hitler died of a rheumatic fever as a young boy, the whole twentieth century might have turned out very, very differently. The rational process plays a role, but only as strings in the webbing of chance and consequence. There is a chance that one person is born and another lives too short a life, a chance that a natural catastrophe destroys a clue to the answer of a critical question, a chance that a natural catastrophe creates a clue, a chance that some document is lost and another found. Trends in the timeline of human destiny are as close to the dull as they are from the fantastic, and they seem to be as chaotic as weather over the high seas.

There is some difficulty with notation printed in ancient documents: no matter how carefully documents are examined, there will always be a certain degree of speculation as to how the notation appeared: Did a scribe introduce something that was not originally part of the manuscript? Did the printer substitute notation from his cold type to keep the easiest spacing? The whole story of what authors were thinking, or how their work influenced others, is the historian's best shot at the most correct story. Sometimes the most expert historians do not agree. The development of human knowledge, like biographies of dead scientists, involves so many intertwined causal scientific, economic, theologic, politic connections that speculation tends to be the best tying knot. There are no tweets telling us what went on in the minds of early contributors to mathematics.

Strange things happen. Good things happen. That's how history works.

The world's most prominent historians of mathematics don't always agree with each other. This is good. It keeps the question open for further study; and isn't that the excitement of history? An organism buried in sediment for hundreds of thousands of years is suddenly exposed by an earthquake. A thousand-year-old scriptural palimpsest in a monastery library is discovered to be a long-lost mathematical treatise. A manuscript that had been preserved by a lava avalanche for hundreds of years surfaces to tell a true story. History gets corrected by unexpected marvels. It happens in every century.

Part 2

Algebra

Going back in time once again, before Indian numerals were brought to Europe.

Significant Initiators

Many of these initiators were either the first or best known for putting symbols into print:

DIOPHANTUS (205 ± 15–290 ± 15). Alexandrian Greek. Mathematician.

> Wrote the *Arithmetica* in ca. 250 AD. First to use symbol for minus (⋔) and unknown (↰).

HYPATIA (ca. 350–370). Greek. Mathematician.

> First notable woman mathematician, and commentator of the *Arithmetica*.

ARYABHATTA (476–550). Indian. Mathematician-astronomer.

> Used letters to represent unknowns.

BRAHMAGUPTA (598–668). Indian. Mathematician-astronomer.

> Possibly the first writer to use zero (a small black dot) as a number (621 AD).
> Wrote the *Brahmasphutasiddhanta* (628), which used abbreviations for squares and square roots and for each of several unknowns occurring in special problems.
> Introduced rules for manipulating negative and positive numbers.

AL-KHWĀRIZMĪ (ca. 780–ca. 850). Persian. Mathematician-astronomer-geographer.

> Scholar in the House of Wisdom. Wrote the *Compendious Book on Calculation by Completion and Balancing* (*Algebra*) (830 AD). Organized rhetorical algebraic expressions according to the various species of forms.

MAESTRO DARDI DI PISA (Jacopo). Italian. Mathematician.

> Unpublished manuscript dated 1344; the *Aliabraa arbibra* earliest manuscript written in the Italian vernacular that exclusively treats algebra.

FRA LUCA BARTOLOMEO DE PACIOLI (1446/7–1517). Italian. Mathematician.

> His treatise on algebra was the first to be printed; gave it the Arabic name *Alghebra e Almucabala* (*Restitution and Comparison*, or *Opposition and Comparison*, or *Resolution and Equation*) (1478).

NICOLAS CHUQUET (1455–1488). French. Mathematician.

> *Triparty en la Science des Nombres* (*Three-part Book on the Science of Numbers*) (ca. 1484). Labeled species of powers as $\mathbb{R}$, $\mathbb{R}^2$,..., and square root as $\mathbb{R}_x$.

JOHANNES WIDMANN (1460–1498). German. Mathematician.

> In his 1489 work, *Behende und hubsche Rechenung auff allen Kauffmanschafft* (*Nimble and Neat Calculation in All Trades*), he introduced + as a symbol for plus.

MICHAEL STIFEL (or Stefleius) (1487–1567). German. Mathematician.

> Published an edition of *Die Coss* in 1553. Used the letters "M" and "D" for multiplication and division, respectively. So 3 ②*D sec* ①*M ter* ② would indicate $\frac{3x^2z^2}{y}$, where *sec* and *ter* stand for second and third unknown.

CHRISTOFF RUDOLFF (1499–1545). German. Textbook author.

> *Die Coss* (1525). Incorporated the symbols ✓, ⩗, and ⩘ for square, cube, and fourth roots, respectively.

GEROLAMO CARDANO (1501–1576). Italian. Physician, mathematician, astrologer.

> Mathematician who in 1545 wrote the *Ars Magna*, solving cubic and quartic equations. Recognized the value of imaginary and complex solutions.

ROBERT RECORDE (ca. 1512–1558). Welsh. Physician and mathematician.

Whetstone of Witte (1557) was widely read, so it introduced the equal symbol (=) to northern European countries.

RAFAEL BOMBELLI (1526–1572). Italian. Mathematician.

Involved with solutions of cubic and quartic equations (1572). Used $\overset{0}{\smile}, \overset{1}{\smile}, \overset{2}{\smile}, \ldots$ to represent the unknown, its square, its cube, and so on.

GUILIELMUS XYLANDER (also Wilhelm Holzmann) (1532–1576). German. Scholar.

A classical scholar. Translator of Euclid's *Elements* and Diophantus's *Arithmetica* into Latin.

FRANÇOIS VIÈTE (1540–1603). French. Mathematician.

Used letters to represent numbers as general objects, and subjected them to the same algebraic reasoning and rules as numbers.

SIMON STEVIN (1548–1620). Flemish. Mathematician and engineer.

In his *L'Arithmetique* (1585), he used the so-called Index Plan for writing exponents—that is, $x^2 - 3x + 2$ would be written as 1②-3①+2⓪.

THOMAS HARRIOT (1560–1621). English. Astronomer, mathematician, ethnographer.

Set polynomials equal to zero, and thereby saw that if a were a root to the polynomial equation degree less than five, then $x - a$ is a factor of the polynomial.

WILLIAM OUGHTRED (1574–1660). English. Mathematician.

Clavis mathematicae (1631). Invented more than one hundred symbols, but less than a dozen survive the seventeenth century. Used × to indicate multiplication and the colon ":" to denote division.

PIERRE HÉRIGONE (1580–1643). French. Mathematician and astronomer.

Cursus mathematicus (1634). Wrote a six-volume algebra text almost entirely in symbols. Invented ⊥ ("is perpendicular to") and ∠ ("angle").

CLAUDE GASPARD BACHET (1581–1638). French. Mathematician, linguist, scholar.

First to translate *Arithmetica* from Greek to Latin (1621).

RENÉ DESCARTES (1596–1650). French. Mathematician, philosopher.

La Géométrie (1637). Used numerical superscripts to mark positive integral exponents of a polynomial. Ranked individual powers numerically. Established the convention of reserving beginning letters of the alphabet for fixed known quantities and latter letters for variables or unknowns.

JOHN WALLIS (1616–1703). English. Mathematician.

Mathesis Universalis and Arithmetica Infinitorum (1655). Used negative exponents and indicated infinity by the symbol ∞.

ISAAC NEWTON (1642–1727). English. Physicist-mathematician, alchemist.

Conceived of unknown variables as Fluents (what we call "dependent variables"), quantities flowing along a curve. Derivatives are denoted as singly dotted forms $\dot{x}, \dot{y}, \dot{z}$, so-called pricked letters.

GOTTFRIED LEIBNIZ (1646–1716). German. Mathematician, philosopher.

Understood the limits and conceptional powers of symbols. Made symbols a priority in his attempts at clear writing. Invented the proper symbols for the differential and integral calculus.

LEONARD EULER (1707–1783). Swiss. Physicist, mathematician.

Represented $\sqrt{-1}$ as i in his Recueil des pieces qui ont remporte les pris de l'academie royale des sciences (1777).

WILLIAM JONES (1746–1794). Welsh. Philologist, ancient India scholar.

Introduced the Greek letter π.

GUSTAVE-PETER LEJEUNE DIRICHLET (1805–1859). German. Mathematician.

Introduced the modern function concept.

WILLIAM ROWAND HAMILTON (1805–1865). Irish. Physicist-mathematician.

Introduced the "quaternions," a new number system in four dimensions that contained the complex numbers.

Chapter 9

Sans Symbols

Many years ago, I had a few rare moments of being permitted to flip through the oldest surviving copy of Euclid's *Elements*, MS D'Orville 301. It was the privilege of a favored few, a privilege no easier than getting permission to visit the queen in her drawing room. First I had to obtain a reference from a respected professor of mathematics. Perhaps it wasn't fully necessary to have it from a knighted professor, but his is what I got. Then, on the day of appointment, a man greeted me in the lobby outside the Special Collections room of the Bodleian Library at Oxford. The gaunt man with a Lincoln-looking face, bushy eyebrows, and sunken cheeks escorted me to a room where he administered an oath.

> *Do fidem me nullum librum vel instrumentum…*
>   I hereby undertake not to remove from the Library, nor to mark, deface, or injure in any way, any volume, document or other object belonging to it or in its custody; not to bring into the Library, or kindle therein, any fire or flame, and not to smoke in the Library; and I promise to obey all rules of the Library.

There, I pledged to respect the property of the Bodleian and agreed to not do things on a long list of things I would not do—neither to use pen nor camera, fire or flame. Next, I was presented with a pair of white gloves and asked to sign the Euclid MS D'Orville 301 guest book with a special pen. I must have hesitated oddly as I glanced at the page I had just signed, for I unexpectedly realized that my signature might be on that page for the next thousand years, just twelve lines below Isaac Newton's.

The gaunt man abruptly reclaimed the pen, and, with a face as serious as Abe Lincoln's, warned, "Under no circumstances do you touch the folio ungloved!"

I was left alone in the room with this magnificent document. I cannot tell you how electrified and privileged I felt to be alone in a room with such an ancient manuscript. I was a monk in a medieval monastery, a count in my Bohemian library, Newton pondering over the question of why there were no symbols in the codex. Alone in that room I felt a spiritual connection to all the past scholars, scribes, mathematicians of the last millennium, and especially to Stephan the Clerk, who in 888 AD laboriously copied the work onto parchment for Arethas of Patras. My white-gloved fingers delicately turned the pages of MS D'Orville 301. (See figure 9.1.)

FIGURE 9.1. Illustration of Big B at Bodleian. Source: Clay Mathematics Institute Historical Archive, http://www.claymath.org/library/historical/euclid/images/euclid_1_48.jpg. The Bodleian Libraries, University of Oxford, MS. D'Orville 301, fols. 31v-32r.

Of course, there were no mathematical symbols other than letters to stand for points and double letters to stand for lines and triple letters to stand for angles. And, of course, there were whole and rational numbers, symbolized by the Greek sequential alphabetic system. But I did not expect and could not find any symbols for addition, multiplication, or equality, other than those scribbled in the margins by so many of those dead readers who must have signed the same oath that I had. Margins were filled with jottings of Indian numerals, along with geometric doodling and even algebraic equations that I fancied had been left by Newton.

Today, we may browse that same text online, no references, and no gloves needed. Thanks to the Clay Mathematical Institute and the Bodleian Library and Rarebookroom.org, the entire original MS D'Orville 301 may be viewed online by anyone. Moreover, each book contains a Greek index keyed to the manuscript images along with English translation.[1]

MS D'Orville 301 shows how to prove simple identities, such as $(a + b)^2 = a^2 + b^2 + 2ab$; however, you will not find any algebraic symbols indicating powers or plus or minus in Euclid's work because his work was geometrical and entirely rhetorical. Even the first printed edition of the *Elements* contained no symbols. Book 2, as translated by Sir Thomas Heath, puts proposition 7 this way:

> If a straight line be cut at random, the square on the whole is equal to the squares on the segments and twice the rectangle contained by the segments.[2]

It is a geometric statement, yet we might see it as our familiar equation

$$(a + b)^2 = a^2 + b^2 + 2ab.$$

Mathematics was not always what it is today. Its rigor did not always count on finite collections of well-formed statements linking back, by rules of logic, to elementary assumptions. Our Western mathematics inherited a fortune in concrete applications from a thousand years of Babylonian and Egyptian calculations, when the concept of mathematical proof was far more relaxed and casual. Persuasion was the aim, not rigor: rigor would slowly develop over the next three hundred years, before Euclid and the Alexandrian school structured the idea of proof based on elementary assumptions.

One of the earliest extant histories of geometry is Proclus's *A Commentary on the First Book of Euclid's Elements*. Proclus was a philosopher and historian who summarized an earlier history by Eudemus of Rhodes. Several historians before the fifth century, from Proclus to Plutarch, tell us that the sixth century BC philosopher

Thales of Miletus introduced a new intellectual marvel to Greek philosophy: abstract geometry.[3]

Little is known of the father of geometry. Even the birth and death dates of Euclid are far from certain. He may have been the compiler and organizer of the great tome called the *Elements*, the 300 BC textbook that encapsulated almost all known mathematics of the time, but certainly was not the sole mathematician responsible for its magnificent proofs. We say, "Euclid proved that…," meaning only that such-and-such a theorem can be found in the thirteen books of the *Elements*. More than likely, this fellow Euclid learned many of the theorems from others connected with Plato's Academy, folks such as Eudoxus and Theaetetus. Learning theorems could not have meant proving them without some axiomatic logical approach. So, though certain geometric theorems may have been sensed as intuitively true by virtue of clever credible arguments, undoubtedly they were less persuasive than Euclid's brilliant organization scheme of building proofs from axioms or self-evident truths obeying indisputable logic. The *Elements* gave mathematics its fundamental nature, its first model of *proof*.

The Pythagorean theorem was known for hundreds of years before Euclid's *Elements* came into being. The Egyptians knew it; the Chinese knew it; the Indians knew it; and certainly the Pythagoreans knew it. But its proof beyond doubt was established at the end of the first book of the *Elements*, after 46 other propositions.

We do know that this fellow Euclid was active in Alexandria shortly after Alexander the Great founded the city in 331 BC, leaving his general Ptolemy I to reign and Dinocrates to architecturally plan the city. The plan was to lay the city out in a grid of straight-line avenues and perpendicular streets. By Euclid's time, the city was already in its energetic days of growth. Two great libraries were already housing copies of books confiscated from ships entering the port. Theaters were performing tragedies, and schools of philosophy were burgeoning.

For five hundred years after Euclid's time, Alexandria continued to be the center for learning and scholarship in mathematics, science, and medicine. By Diophantus's lifetime in the second century, the city was still a marvel. Its wide boulevard and side streets were paved with stone and dimly lit by torches at night, better lit than European cities would be for the next two thousand years. It still had those limestone colonnades that extended from one side of the city wall to the other. It still had its parks and its monuments to Cleopatra. Perfumes and parchment were still part of a thriving manufacturing industry, along with glassworks and alabaster carving. It still had its temples and synagogues. It still had its street vendors, performers, moneylenders, and prostitutes. It was "a mood-altering city of extreme sensuality and high intellectualism, the Paris of the ancient world."[4] No wonder so many mathematicians came to work there.

Diophantus was probably born there, though we don't really know for sure. We can say for sure that his masterwork, which was forgotten through the entire period of the dark ages, had a profound influence on the development of algebra when it was rediscovered in the sixteenth century.

Now these were still years when reading meant reading aloud. Silent reading did not yet exist, not even in public. Reading was a concentration effort that required balancing the scroll and rolling it.[5] By Diophantus's time, words were already separated. But earlier manuscripts were written without word, sentence, or paragraph breaks, and without punctuation. Reading was difficult and considered a masterful accomplishment. Reading and writing was done mostly in the morning hours when there was light, and when the heat of the day had not begun. Diophantus would have worked on parchment using a reed of sea rush.

He did use symbols for powers and unknowns, if we can trust the scribes and translators that copied and translated his work. For unknowns, he would use a symbol that resembles the Greek sigma ς used at the end of a word, although bigger and more tilted than the sigmas appearing within the text.[6] It may have been the cursive contraction of the first two letters of the Greek word for number. It may not appear that way to us, but the renowned mathematical classicist Sir Thomas Heath believed that it may have evolved through a sequence of progressively sloppy scribe copying. Diophantus also concocted a minus sign, something that looks like an arrow facing upward (sometimes facing downward). But this too was a contraction—quite possibly the first two letters of the word for minus, one nested in the other.[7] Or, it might have simply been that some scribe in a later century had the idea on his own to use these symbols as abbreviations.

What would mathematics be like if it were entirely rhetorical, without symbols, or without its abundance of cleverly designed symbols? Here is a passage in al-Khwārizmī's *Algebra* (ca. 830 AD), where even the numbers in the text are expressed as words:

> What must be the amount of a square, which when twenty-one dirhems are added to it, becomes equal to the equivalent of ten roots of that square?[8]

We would write this simply as $x^2 + 21 = 10x$.

The solution without symbols reads as follows:

> Halve the number of roots; the moiety is five. Multiply this by itself; the product is twenty-five. Subtract from this the twenty-one which are connected with the square; the remainder is four. Extract its root; it is two. Subtract this from the moiety of the roots, which is five; the remainder is three. This is the root of the square which you required and the square is nine...

Need we go on?

The question comes from a slightly more practical question:

> I have divided ten into two parts. Next I multiplied one of them by the other, and twenty-one resulted. Then you now know that one of the two sections of ten is a thing.[9]

The language of the solution, as al-Khwārizmī wrote it, provides us with a procedure that seems to be specific to the question. There may be a routine method, some algorithm lurking behind the phrasing, but it may take some work to express the process. On the other hand, the symbolic algebraic process extracts the answers to many questions of that genre. In modern symbolic terms, the solution works out this way: Ten is divided into two parts, where one part could possibly be bigger than the other. So the two parts may be represented as x and $10 - x$. The product of those two parts must equal 21. Hence, $x(10 - x) = 21$. This expands to become the quadratic equation $x^2 - 10x + 21 = 0$, whose solution is $x = 3$ or $x = 7$.[10]

However, keep in mind that the problem was not likely worked out rhetorically. It would have been first worked out on some sort of a dust board, and afterward composed rhetorically for text presentation. Moreover, the question of how to divide 10 in two parts so the product of the two parts is 21 has a simple answer that could even be worked out in the mind: 21 has only two factors, 3 and 7, which, when added together gives 10. It is also a standard Babylonian geometry problem: list the two parts x and y, and look at the problem as one where x and y are two sides of a rectangle whose sum is 10 and whose area is 21. That geometry problem is reflected in the algebra problem that considers two equations to be solved:

$$\begin{cases} x + y = 10 \\ xy = 21 \end{cases}$$

Solve the first for y and plug it into the second to get $x^2 - 10x + 21 = 0$.

Al-Khwārizmī's proofs are geometric, not algebraic in the modern sense of what we call algebra. That is not surprising, as there are no algebraic proofs in Arabic mathematical writing of al-Khwārizmī's time.[11] But, without any explanation, al-Khwārizmī does give us a kind of rhetorical algorithmic reduction from the question to its answer. By rhetorical algebra that is very hard to understand, he tells us how to proceed. Yet there are no symbols, not one, not even numeral symbols. Jens Høyrup tells us that "algebraic proofs for the solution of the basic equations are absent from the entire Arabic tradition. ... We should expect no direct connection between the existence of an algebraic symbolism and the creation of the kind of reasoning it seems with hindsight to make possible."[12]

We should not be fooled into thinking that the symbolic form of a rhetorical statement is just convenient shorthand. It is shorthand, all right; but more than that, it helps the mind transcend all the ambiguities and misinterpretations dragged along by written words in natural language. Even more, this symbolism permits the mind to lift particular statements to their general form. By Descartes's time, equations were written in almost completely modern symbolic form: the symbol had finally arrived to—as Tobias Dantzig put it—"liberate algebra from the slavery of the word."[13]

Chapter 10

Diophantus's *Arithmetica*

When you get to know them, equations are actually rather friendly.[1]
—Ian Stewart

The earliest works of the ancients that could be called algebra date back to the early Pythagoreans, or at least perhaps the Pythagorean Thymaridas of Paros, who, according to the Syrian philosopher Iamblichus, gave a rule for solving a certain set of n simple simultaneous equations in n unknowns. For three unknowns, the rule simplifies to: *Given a sum of three quantities and also the sums of every pair containing one of those specified quantities, then that specified quantity is equal to the difference between the sums of those pairs and the total sum of the original three quantities.*

We, with our modern symbolic language, might say it more simply this way:

If, simultaneously,

$$\left\{ \begin{aligned} x + y + z &= a \\ x + y &= b \\ x + z &= c \end{aligned} \right\},$$

then $x = b + c - a$. For instance, if simultaneously,

$$\left\{ \begin{aligned} x + y + z &= 3 \\ x + y &= 2 \\ x + z &= 4 \end{aligned} \right\},$$

then $x = 2 + 4 - 3 = 3$. This is an easy substitution process, but also, in essence, what has been known since the nineteenth century as Cramer's rule. In Thymaridas's time, it was called "the flower of Thymaridas." If we continue to find the other unknowns, we get $y = -1$ and $z = 1$. Although the solution that uses $y = -1$ works, it would have been considered absurd, because -1 is a negative quantity. Fractions and rational numbers would have been fine; however, before the sixteenth century, negative numbers—which had been perfectly acceptable as debts—would not have been accepted as genuine numbers in Europe.[2]

The ninth-century version of Euclid's *Elements* (MS D'Orville 301), books II, V, and VII, unwittingly deal with algebra through the language of geometry—that is, magnitudes pictured as line segments to give, for instance, a solution to the quadratic equation $x^2 + ax = b^2$ (in our modern notation), where a and b are positive numbers. We, in the twenty-first century, mean that there are two numbers that can replace x and balance the equation; calculating with one of those numbers will make the left side the same as the right side. But for anyone living before the fifteenth century, the solution would be found without symbols along with a concrete view of what *things* would be acceptable as numbers.

The equation $x^2 + 3x = 4$ has two solutions ($x = 1$ and $x = -4$); however, only one is positive, and therefore, only one was acceptable as a number. Such an equation may be solved geometrically; in fact, it would have been inspired by a practical geometric problem—say, to find the width of a rectangle whose length is three *stadion* (approximately 607 yards) more than its width and whose area is equal to four square *stadion*. In such a case, for such a geometric problem, the negative number -4 that satisfies $x^2 + 3x = 4$ seems to be inapplicable to a rectangle that must have positive dimensions. Fifteenth-century mathematicians could not have known that the geometry itself held the answer to why that negative number satisfied the equation. The equation itself gave something that the geometry was not picking up, even though that negative number solution represented something quite real in the geometry.

Modern mathematics comes from essentially three roots: algebra, geometry, and analysis, with logic as the earth of all three. Those roots are entangled and knotted in the undergrowth where adhesions make it difficult to distinguish one root from another: we now have algebraic geometry, a relatively new branch of mathematics that combines techniques of abstract algebra with those of geometry; geometric analysis, a discipline that uses geometrical methods to study partial differential equations; and analytic number theory, a branch of number theory that uses analysis to solve integral problems. Yet, fundamentally, at mathematics' very old undergrowth root level, we find algebra, geometry, and analysis.

Symbolic modern mathematics, at its most rudimentary stage, can be traced back to Diophantus's *Arithmetica*. We should be warned that the original text does not survive, so any notation found in copies might have been introduced by scribes or translators.

Diophantus wrote after Hypsicles of Alexandria (because Diophantus quotes Hypsicles) and before Theon of Alexandria, the father of Hypatia (because Theon quotes Diophantus). This would put his time on Earth roughly between 120 and 400 AD. There is also a letter from an eleventh-century Byzantine monk claiming that Anatolius, the Bishop of Laodicea (in present-day southwestern Turkey), dedicated a treatise to Diophantus sometime near the year 250 AD. Such a dedication would suggest that Diophantus must have been active not much later than 250 AD.[3]

However, in the 1980s, the distinguished historian of mathematics Wilbur Knorr suspected that a book that had long been attributed to Heron of Alexandria was actually written by Diophantus. Knorr examined the style of the book that was allegedly written by Heron and found that its style closely resembled Diophantus's. He hypothesized that the Bishop of Laodicea's letter must have referred to a different Diophantus. Heron died in 70 AD, so that would put the original *Arithmetica* in the first century rather than in the third.

An epitaphic poem may tell us something about Diophantus's age:

> "Here lies Diophantus," the wonder behold.
> Through art algebraic, the stone tells how old:
> "God gave him his boyhood one-sixth of his life,
> One twelfth more as youth while whiskers grew rife;
> And then yet one-seventh ere marriage begun;
> In five years there came a bounding new son.
> Alas, the dear child of master and sage

After attaining half the measure of his father's life chill fate took him.

After consoling his fate by the science of numbers for four years, he ended his life."

It is an algebra puzzle coming from a seventh century collection of puzzles in the Greek Anthology, *Anthologia Palatina*, written under the name Metrodorus. A solution back then would have required a masterful juggling of the material found in the *Arithmetica*; yet with our symbolic algebra the solution is quickly found.

Following the poem, we find that Diophantus's youth lasted 1/6 of his life. After 1/12 more of his life, he grew a beard. After 1/7 more, he married. Five years later, he had a son, who lived half as long as his father. Diophantus lived four years after his son's death. So, if we let x equal the number of years Diophantus lived and y equal the number of years his son lived, then we know that

$$x = \left(\frac{1}{6} + \frac{1}{12} + \frac{1}{7}\right)x + 5 + y + 4,$$

and that

$$y = \frac{x}{2}.$$

These may be considered to be simultaneous equations in two unknowns, but they quickly reduce to one simple equation in a single unknown. By substituting the y of the second equation into the first, we find that Diophantus died at age 84. How easy was that?

The *Anthologia Palatina* contained 46 epigrammatical puzzles, many of which were algebraic in nature leading to simple simultaneous equations coming from a tradition of problems of dividing apples among some number of persons. Such algebraic puzzles, written without a single symbol, date back to before the fifth century BC. One, for example, asks for the number of apples that can be divided between six persons so that the first receives one-third; the second receives one-eighth; the third receives one-fourth; the fourth receives one-fifth; the fifth receives 10 apples; and the sixth receives just 1.

We see this as just the question, what is x, if

$$\frac{1}{3}x + \frac{1}{8}x + \frac{1}{4}x + \frac{1}{5}x + 10 + 1 = x?$$

By the tools of our symbolic algebra, we manipulate the equation: we add the like terms together, subtract x from both sides, and quickly get the answer $x = 120$ apples.

As a rule, the translations of Greek texts to Syriac to Arabic to Latin had to go through several stages, each adding a fair share of inaccuracy. The intermediate translations went through Persian, Syriac, Arabic, Aramaic, and other languages. The Arabs were interested mostly in science, mathematics, mechanics, and philosophy—Apollonius, Philo, Archimedes, Heron, Plato, Aristotle, and Theophrastus. By the middle of the ninth century, in Baghdad, in Byzantium, and elsewhere along the eastern Mediterranean, there was a growing interest in scholarship with increasing calls for translations. In Baghdad, there was Hunain ibn Ishaq, a seventeen-year-old polyglot who founded a school for translators. Hunain circulated his suspicion that Greek manuscripts were scattered all over the Islamic world, and personally led expeditions to find them in Mesopotamia, Syria, and Alexandria. He was contemptuous of earlier translators, whom he claimed were either completely incompetent or were working from damaged or illegible manuscripts.

Hunain's school was special because his technique for translating was different and right, at least by modern philological standards. His school taught students to scrupulously compare divergent manuscripts whenever and wherever they could be found. "Thanks to the scholarship of Hunain and his associates many Greek texts survive as high quality Arabic translations."[4]

Before the fourth century, words in books were written in uncial characters (capital letters). Although there were some experiments with lowercase script during the next few centuries, surprisingly little had changed before the founding of Hunain's school. Uncial writing had the critical disadvantage of being too slow and too large to write; the amount of text on a page was limited. To cut down on expensive writing material, minuscule script (lowercase letters that had been used for letters and official documents) replaced the uncial script. The new script made copying easier and cheaper. Books could be scribed far more quickly, but with the hindering nuisance of ambiguous scripts in constant need of interpretative decisions.

After the Arab conquest of Egypt in 641 AD, the demand for parchment increased considerably, even though there was not much interest in literature at that time. Papyrus plantations were depleted. Writing materials were no longer cheap and no longer readily available. But by 850 AD, coincident with a revival of scholarship—or very likely assisted by that revival—manuscript writing changed both in appearance and in production.

Then, in 751, during the battle of Talas, in which the Arabs halted the Chinese western expansion into Kazakhstan, two Chinese prisoners of war were taken at Samarkand. The Kazakhstani Arabs learned the process of papermaking from those two Chinese soldiers. Paper brought the cost of writing down to affordable. So, in the ninth century, the old uncial texts could be transliterated into the new minuscule script to preserve the best of Greek literature. All later copies of ancient Greek texts are derived from one or more uncial script predecessors written on papyri; almost all are derived from their ninth-century exemplars.

Unfortunately, mistakes happen through transliteration. Letters are confused and misread. Many errors from the Greek appear to be usually derived from the same sources of ninth-century manuscript copies. After minuscule copies were made from uncial sources, the originals were discarded, and the minuscule copies became the sources of all further copies. So, many texts survived in one copy only. As for Diophantus's *Arithmetica*, which originally consisted of thirteen books according to its preface, only six and part of a seventh survive.

A quick glance of the *Arithmetica* gives hints of algebraic character. This is why some historians in the past have suggested that algebra began with Diophantus. More careful glances reveal the brilliance of the work as well as the crudeness of its notation. The book teaches us how to solve specific equations of the first and second degree, yet its notation looks as if it is composed of abbreviations of unknowns and powers that are carried into the calculations toward solutions.

Posing and answering difficult questions about squares, cubes, and other general properties of numbers to someone by the name of Dionysius (no relation to the god), the work defines letter-based names for squares and cubes, and labels the unknown quantity ἀριθμός, meaning "the number." Very soon into the work, Diophantus uses the symbol ↳, as if the full word ἀριθμός is a bother.

For two hundred years, scholars have been questioning the origins of this symbol ↳.[5] Some thought it to be ς the alternate to the Greek letter sigma that was used only at the end of a written word. The thought was that Diophantus knew there would be no confusion between ς and a number; letters of the Greek alphabet had a numerical equivalent, but that last letter ς (the alternate sigma) was never considered a number under the Greek numerical system. That was the argument favoring ↳ as just a large tilted ς. There is another argument, however, that favors the idea that ↳ represents a shorthand contraction of the first two letters (the first syllable) ἀρ of the word ἀριθμός, and is in no way an algebraic symbol by our definition of symbol.

All respectable arguments have worthy reasons. At the turn of the twentieth century, the eminent mathematics history scholar Sir Thomas Heath gave a persuasive rationale for his belief that ↳ was neither the final sigma nor some sort of hieroglyph, but rather a deformation of the two first letters of the word ἀριθμός. He reasoned that it establishes a "uniformity between the different abbreviations used by Diophantus. It would show him to have proceeded on one invariable principle in fixing those abbreviations which we should naturally have expected to be parallel."[6] The letters μ, δ and κ correspond to the first letters of the Greek words they are meant to represent, the monad (our unknown x), the square, and the cube. Heath argued that these letters also could be confused with their corresponding numerical equivalents—namely, 40, 4, and 20. To avoid such a confusion, Diophantus must have added the second letter of each of the Greek words, μονάδων, δύναμις, and κύβος. But then μο, δύ, and κύ could be confused with the numerical equivalents 4,070, 4,400, and 20,400, respectively.[7] To avoid *that* confusion, the second letter of each word was superscripted. The abbreviations became $μ^o$, $δ^ύ$, and $κ^ύ$. Applied to the word ἀριθμός, the abbreviation would be ἀᵖ. Occasionally, other symbols appeared, such as μᵒ, which stood for an indefinite number.

How does Heath get from ἀᵖ to ↳? For the moment, let's ignore the fact that in Diophantus's time fractions were normally written with the denominator as a superscript of the numerator, which means that ἀᵖ might be confused with the representation for 1/100.

Scribes were not always careful with their cursive. The scribe, working quickly through long hours—sometimes by the light of a cloudy window, sometimes by dim candlelight—might turn the cursive form of the pair of Greek letters αρ into a graphic image similar to ↳ or Ꙅ, for both s-like shapes were used in later translations. Heath appealed to the eminent nineteenth-century philologist Viktor Emil Gardthausen, who argued that cursive writing in ancient manuscripts went through stages. The pair of Greek letters ἀᵖ morphed to become ᵁᴾ. This may have been Diophantus's shorthand for the word ἀριθμος. Then, according to Heath, after much copying and recopying, successive generations of scribes would no longer see the marking as two letters, but rather as some obscure minuscule form to be scribed as seen.

The job of a scribe was to copy, not to copyedit, and surely not to enhance or alter content. Scribes were either monks or professionals for hire, who often had no idea what they were actually copying, especially when the books they were copying were scientific or mathematical. With the job came the perk of being left alone for months, sometimes years at a time. Often the author had been dead for many years or centuries, so there was no one with authority available to check for mistakes. Left on their own, those scribes embellished, added, deleted, and made mistakes. The more celebrated texts, such as the *Arithmetica*, were already copies, so mistakes tended to compound seriously enough to infuriate medievalists as they tried to distinguish author from copyist.

As plausible as Heath's argument sounds to me, it had been challenged by many scholars. D'Arcy Wentworth Thompson, the early-twentieth-century Scottish mathematical biologist, had his own theory of how $\hookleftarrow$ came about.[8] The symbol was generally written with appended inflections as ς', or $\varsigma^{o\bar{u}}$, or (in the plural case) $\varsigma\varsigma^{oi'}$, which suggest that the symbol is supposed to be part of a word. The nineteenth-century mathematics historian James Gow wrote a friendly essay soon after Heath's idea surfaced in the early-twentieth-century scholarly world of mathematical history. Gow believed that $\hookleftarrow$ is neither a contraction of the first two letters nor the final sigma in the word ἀριθμος. He dismissed the final sigma opinion by arguing that ς appears only in cursive Greek, and that cursive Greek did not appear before the eighth century.[9] With doubt that $\hookleftarrow$ comes in some way from corrupted shorthand, he entertained the thought that it may have come from Indian or Babylonian, or hieratic (Egyptian cursive writing) characters. His friend Samuel Birch, the Egyptologist, told him that, in form, ς' is practically identical to the hieratic sign of a papyrus-roll, which also signified an unknown force and also a "heap" (the Egyptian *hau*). Ahmes used a hieratic sign of a papyrus-roll to mean the unknown. He was the scribe of the famous Rhind papyrus, a handbook of practical problems dating from about 1550 BC, "a guide to accurate reckoning of entering into things, knowledge of existing all things," now in the British Museum. All hieratic signs differ slightly in form, and are derived from different hieroglyphic pictures; however, the sign for "a sum-total" also seems to be very close to that of the papyrus-roll.[10] After Gow published his argument, Heath rebutted it.[11] So the whole question remained outstanding.

Diophantus did not have any symbol for "plus." However, here lies another mystery: In 1621, the French mathematician, linguist, and scholar Claude Gaspard Bachet translated Diophantus's *Arithmetica* into Latin. According to that translation, Diophantus clearly tells us (book I, definition IX): "*A wanting* [minus] multiplied by *a wanting* makes *a forthcoming*, and that the *wanting* is denoted by the letter ψ truncated and upside down so as to form ⋔."[12] Often the symbol ⋔ would appear to indicate minus, a true symbol, one that is abstract with no apparent direct association with the written word "minus."

> DEFINITIO IX Minus per minus multiplicatum, producit Plus. At minus per plus multiplicatum, producit minus. Et defectus nota est litera ψ decurtata, & deorsum, sic ⋔.

Translation: Definition IX: Multiplying less than by less than produces more than. Multiplying more than by less than produces less than. And the minus is denoted by the letter ψ cut short and turned upside down.

Here we have the first evidence of the symbol for minus. Diophantus tells us that his symbol for minus ⋔ is the Greek letter ψ with its tail cut and turned upside down. The notation is not always consistent, however; sometimes the symbol is used and sometimes, in the original Greek, the word λείψει (wanting), even on the same page.

The same symbol appears in Heron of Alexandria's *Metrica*, written in the first century AD, which would mean the symbol was used before Diophantus was born. It may have been an abbreviation of the word λείψει, perhaps the first and last letters superimposed, or perhaps some hieratic character.[13] It seems that ⋔ is the only Diophantus mark that may be a true symbol with no direct association with the written word. All other markings in *Arithmetica* seem to be abbreviations. Since all surviving copies were made from one dating back to the thirteenth century, it is hard to know who is responsible for any of the symbols that could have entered along the way.[14]

To indicate a sum of two terms, Diophantus (or a scribe of *Arithmetica*) would simply juxtapose them. For example, the unit $\mu^o \overline{\alpha}$ would be joined to the unknown $\varsigma^{o\overline{v}}$ to represent the polynomial $x + 1$ in syncoptic notation as $\varsigma^{o\overline{v}}\mu^o\overline{\alpha}$, or more simply as $\varsigma\mu^o\overline{\alpha}$. Yet he was able to simplify equations by a transposition process of adding, subtracting, and collecting like quantities, just as we do, all done rhetorically and without rules of procedure. He assumed that we must know the rules from somewhere else, perhaps from another book or from another teacher. So whenever he miraculously arrived at a solution, he stopped. Incidentally, we still use the juxtaposition idea when writing fractions; the mixed number $2\frac{1}{2}$ means $2 + \frac{1}{2}$.

The Copies of *Arithmetica*

Only six of the thirteen books survive as copies of the original in Greek. Four of the thirteen in Arabic (books IV through VII) were discovered as recently as 1968.[15] Almost every current commentary on the *Arithmetica* comes from the Bachet's Latin translation of the Parsinius 2379 manuscript that was scribed by Ioannes Hydruntius sometime after 1545. That manuscript, now in the Paris Bibliotèque Nationale, was the first edition that contained the Greek text.[16] Tracing the original of *Arithmetica* is difficult. The earliest manuscript dates no further back than the badly preserved thirteenth-century manuscript *Matritensis 48*, now at the Madrid Biblioteca Nacional.

The great libraries of Europe began as small suites of rooms. Some of the earliest universities were established in small cities of Italy before the middle of the thirteenth century—Bologna, Florence, Naples, Padua, Pavia, Perugia, Pisa, Rome, and Siena. These became the scholarship centers of Italy long before the existence of the Vatican Library. For many of these cities, a university (*universitas*) was still just a society of students bound by the scholastic interests of individual teachers. It had no physical establishment. Students of family wealth came from all over Europe to the small towns of Italy to study under the common language, Latin. They would pay their teachers directly.[17] Free, liberated from the working class, they were the students of the *liberal arts*.

In 1463, the German mathematician and astronomer Johannes Müller, who went by his Latin name Regiomontanus, lectured at the University of Padua, a school that had already been established for over two hundred years. In connection with those lectures, he gave a talk that purported to introduce all the mathematical sciences. "No one," he reported, "has yet translated from the Greek into Latin the fine thirteen Books of Diophantus, in which the very flower of the whole of arithmetic lies hid, the *ars rei et census* which today they call by the Arabic name of Algebra."[18] It may have been the first time a European writer mentioned a work by Diophantus.[19] Then, in a letter to the Italian mathematician Giovanni Bianchini, he wrote that he had found in Venice "Diofantus, a Greek arithmetician who has not yet been translated into

Latin."[20] Nobody seems to know for sure how Regiomontanus found that copy of *Arithmetica*. Sometime near 1620, Bachet claimed that Cardinal Perron possessed a manuscript containing the complete thirteen books of Diophantus. According to Perron, it was loaned to a friend, who died before it could be recovered.[21]

Two centuries before the Ottoman siege of Constantinople in 1453, a fire in that city's great library destroyed over a hundred thousand books. Yet a few years later, the library managed to dedicate its resources to translating Greek and Aramaic to Arabic and pay hundreds of scribes to transfer disintegrating ancient papyri texts to parchment. Somehow, copies that were once part of the great Constantinople Library migrated west as spoils of war, and eventually wound up in private hands, in the growing university libraries all over Europe as well as in the Vatican.

Then, in 1448, Pope Nicholas V created a public library space in the papal palace. It began with a suite of frescoed rooms with large windows. Books that the pope deemed of great consequence or of great illuminated beauty were chained to benches. That library space became its own thing of beauty. By the time of Nicholas V's death in 1455, its collection was well over one thousand books. The first Vatican librarian, Bartolomeo Platina, who was appointed by Pope Sixtus IV in 1475, handwrote a catalogue of 3,500 entries, the largest collection of books in Europe.

The books were mostly theological. By the end of Platina's six-year tenure, however, the collection of secular works in Greek and Latin grew to become the most important scholarship center for classical works in the Western world, with thousands of illuminated manuscripts on art, music, philosophy, theology, history of the Roman church, science, and mathematics that were bought or looted from kingdoms and empires as far east as China. We now know that at that time the Vatican Library possessed at least two copies of Diophantus's writings.

The sixteenth-century German scholar Guilielmus Xylander tells us that he came across a copy of *Arithmetica* in October 1571, when he was in Wittenberg talking to a couple of mathematicians who already had possession of a few pages of the *Arithmetica* manuscript that belonged to one Andreas Dudicius. Before Xylander left Wittenberg for Leipzig, he copied one problem and its solution to show to Simon Simonius Lucensis, a professor at Leipzig, who wrote back to Dudicius for the manuscript.

The next younger manuscript is the fifteenth-century (Vat. gr. 191) copy of *Matritensis 48* at the Biblioteca Apostolica Vaticana. Three centuries later, the French mathematician and historian of mathematics Paul Tannery compiled and organized a list of twenty-three copies of the *Arithmetica* from the thirteenth to the sixteenth century.[22]

Hypatia, who lived in the fifth century, had a copy that was later lost. There were references to an eighth-century or ninth-century copy as well. Close to a thousand years had passed between the time Diophantus wrote his *Arithmetica* and the time Matritensis 48 was written. Copy after copy, from Greek to Arabic, to Aramaic, back to Greek must have incurred not only errors but also extras. Couldn't the abbreviations we attribute to the original have found their way into one of the copies along the way?

It is very difficult to follow the notation from one copy to another. Examining three translations, we find wild differences: the English of Heath, the Latin of Xylander and the Latin/Greek side-by-side of Bachet.[23] Heath transcribed the symbols into a form that is not recognizable in both Xylander and Bachet, and yet almost all the popular literature use Heath's transcriptions. In the Madrid manuscript (Matritensis 48), the unknown appears as ҷ, very much like a vertical and horizontal reflection of the Latin letter "h." In the fifteenth-century Venice manuscript (Marcianus 308), that same marking appears as S, and in the Bodleian as $\overset{\prime}{\underset{}{S}}$.[24] Heath argued that all those symbols were simply corruptions of an abbreviation. Such a hypothesis would be more consistent with how the symbols μ, δ, and κ came to denote square (power) and cube from μονάδων, δύναμις, and κύβος, respectively.[25]

Even Bachet's translation gives several notations for the unknown (our x). In his second definition, it appears as something that looks like the Greek letter ς. At times, it has an accent ς́; at other times, it appears with a superscript ς″; still at other times, with a super-superscript ς$^{o'}$. These are all shorthand representations of what Diophantus calls ὁ ἀριθμός, "the number." Sometimes we find the symbol written as ς^{οι}; at other times, ς^{όν}. These variations reflect the grammatical or semantic form of how the indirect object "number" is used, for they reflect the word endings of the various possibilities for ἀριθμός (ἀριθμοί or ἀριθμόν), depending on their grammatical structure within the sentence. A double sigma indicates plural; on the same page, we may see ςς^{ούς}, or ςς^{ιον}, or ςς^{οις}, again depending on the grammar.

The letter ς also appears in a work of the Platonic philosopher Theon of Smyrna, who lived in the early part of the second century. So Theon may have been the first to think of abbreviating words in mathematics.

The polynomial in Parsinius 2379 translates to $9Q+14-9N$ in Bachet's notation (where Q represents x^2 and N represents x)[26] and to $x^2-9x+14$ in ours.[27] Note the ςξ. Here Diophantus is using the plural of the object, because he has combined nine negatives into one term. Note that Diophantus's coefficients (that is, ϑ= 9 and ιδ= 14) are written after the species demarcations (that is, ςξϑ). In our notation, ςξϑ would mean $9x$, where the x as a double sigma means the plural, 9 x's. In other words, $1x$ would be written as either ςα, or simply ς, whereas $2x$ would be written as ςξβ. (See table 10.1.)

Table 10.1. Listing of Diophantus's Notation

μ^ō (units)—for example, μ^{ōε} means 5 units

⋔ (minus)

ι^σ (equals)—probably from the first two letters of the word ισος, which means "equal"

↳ (unknown) x

δ^γ (square) x^2

κ^γ (cube) x^3

δ^γδ (square-square) x^4

δκ^γ (square-cube) x^5

κκ^γ (cube-cube) x^6

Note: In the classical commentaries, these symbols are capitalized. See Diophanti Alexandrini, *Opera Omnia.* Also see Heath, *A History of Greek Mathematics*, 448.

All this may indicate that Diophantus was surely representing the unknown, not as a symbol that would be visibly devoid of the word (yet conceptually connected to the word), but rather as a mere abbreviation. However, the nineteenth-century mathematics historian Paul Tannery claimed that the ancient manuscripts before Byzantine times did not use these different grammatical cases and that it was likely that later copyists took it upon themselves to include the case endings as abbreviations. If Diophantus used a different notation for different grammatical case endings for the unknown ἀριθμός, then why did he not do the same for other symbols?[28] Adding to the confusion of hypotheses, Heath was suspicious of the argument that Diophantus actually used a final sigma as the abbreviation for the unknown. His reasoning is that that final sigma was a later addition to the Greek alphabet. That suspicion would have been strengthened by Bachet's translation of Definition IX, which suggests that it has nothing to do with the final sigma.

Translated to our modern notation, the Diophantine polynomial κ$^Υ\overline{γ}$⋔δ$^Υ\overline{β}$ς$\overline{α}$μ$^ο\overline{α}$ becomes $3x^3 - 2x^2 + x + 1$:

$$\underbrace{κ^Υ\overline{γ}}_{3x^3} ⋔ \underbrace{δ^Υ\overline{β}}_{-\ 2x^2} \underbrace{ς\overline{α}}_{x} \underbrace{μ^ο\overline{α}}_{1}$$

Diophantus used the mark **x** to write reciprocals. To write $\frac{1}{x}$, he would write ς^x.[29] But division was marked by the words ἐν μορίῳ, which means "in sharing." So in our notation, δ^υ τμ^ο αψκε ἐν μορίῳ δ^υ δαμ^ο πδ δ^υ μ is written as:

$$\frac{300x^2 + 1725}{x^4 - 40x^2 + 84}.$$

ϛ´ μ´ δ´. ἐν μορίῳ τῆς ν.[30]

Of course, from our modern standpoint, we see Diophantus's notation as . . . well . . . not so difficult to understand, but rather extremely difficult to algebraically manage. It is a cumbersome notation. Even Diophantus tells us, "You will think it hard before you get thoroughly acquainted with it."[31] Since there was no sign for addition, it was necessary to group all the negative terms together after the sign for subtraction. Moreover, his notation gave no signal to the mind that x and x^2 are of the same number species.

We might say that Diophantus's notation is terribly awkward and acutely difficult to process compared with what we have today, and be amazed that he could do any kind of mathematics under the circumstances. We might think that such notation must have hindered clear algebraic thinking. Perhaps, but routine and familiarity are the tailwinds of conception. The problem for us is that his notation clothes everything in the same way and does not solidly distinguish operational symbols, such as powers or summation, from numbers or indeterminate objects. Without breaks between the powers—that is, by pluses and minuses—the mind might have to work harder to comprehend the algebra.[32]

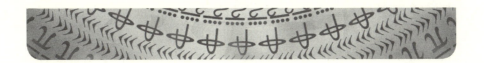

Chapter 11

The Great Art

The art of algebra may have come from the Greeks or from the Hindus. However, the Brahmins of northern India had some idea of algebra long before the Arabians learned it, contributed to it and brought that art to Spain in the late eleventh century. The Indian mathematician Brahmagupta wrote the *Brahmasphutasiddhanta* in 1,008 metered verses "for the pleasure of good mathematicians and astronomers." Completed in 628, it not only advanced the mathematical role of zero but also introduced rules for manipulating negative and positive numbers, methods for computing square roots, and systematic methods of solving linear and limited types of quadratic equations.[1]

The tenth-century *Meadows of Gold* encountered in chapter 7 reported on an older book on science and astrology called the *Sindhind* (*The Revolving Ages*), a calendar book of astronomical tables recording the positions of the sun, the moon, and known planets, along with astrological data, and a table of trigonometric signs. It was the encyclopedia of everything the Hindus knew about arithmetic, astronomy, and all other sciences.

Al-Khwārizmī read through the *Brahmasphutmasiddhanta*, and soon turned his energy to writing an Arabic version, the *Zīj al-Sindhind*, an astronomical treatise based on the methods of the Sindhi and Hindus of India and completed sometime before 825. In working with particular mathematical questions, he became fascinated with the methods of finding missing quantities that were originally rhetorical and in some cases abbreviated. Five years later, al-Khwārizmī published a book titled *Al-Kitab al-mukhtasar fi hi sab al-gabr wa'l-muqabala*, which translates roughly as "The Compendious Book on Calculation by Completion and Balancing."[2] Note the word *al-jabr* in the Arabic title; *al-jabr* is sometimes translated as "restoration" or "completion." In fact, the term comes from the Arabic verb "to set," as in "to set a bone." Like the fate of many other handwritten manuscripts before the early days of Gutenberg's movable type in the mid-fifteenth century, the only surviving copies of al-Khwārizmī's *Algebra* date back no further than the fourteenth century. Aside from minor fragments, however, three complete copies do survive.[3]

Fra Luca Bartolomeo de Pacioli (1446/7–1517), whose treatise on algebra was the first to be printed, gave it the Arabic name *Alghebra e Almucabala* (*Restitution and Comparison*, or *Opposition and Comparison*, or *Resolution and Equation*). Pacioli also called it *L'Arte Magiore: ditta dal vulgo la Regola de la Cosa over Alghebra e Almucabala* (*The Great Art: Commonly, the Rule of the Thing in the Art of Restitution and Comparison*).

Other authors claim it from other Arabic words. Pierre de la Ramée, the sixteenth-century French mathematician dubiously claimed in his *Arithmétique* (1555) that "the name algebra is Syriac, signifying the art and doctrine of an excellent man." He went on to say that there was a certain learned unnamed mathematician who wrote a book for Alexander the Great named *Almucabala*, a book of dark or mysterious things, which was later called *Aljabra*, the doctrine of algebra.[4]

Mysterious, yes. But dark? *Almucabala*, the title of Robert of Chester's Latin translation of al-Khwārizmī's book, does give that onomatopoeic feel of dark and mysterious things to an English-speaking person. Perhaps both names are appropriate for the art.[5] What may have been mysterious to a student of the art in ninth-century Persia is familiar to a student of today. The *Almucabola* gives this example:[6]

> I divide ten into two parts in such a way that the product of one part multiplied by the other gives 21.

It doesn't ask us to find the two parts. Rather it proceeds to give the method for finding the answer to what we think is the question. (The square brackets are my interpretations.)

Now then we let root $[x]$ represent one part, which we multiply by ten less root $[10 - x]$, representing the other part. The product ten roots less the square $[10x - x^2]$ equals to 21. Complete ten roots by the square [10 by x^2] and add this square $[x^2]$ to 21. This gives ten roots $[10x]$ equal to the square plus 21 $[x^2 + 21]$. Take one-half of the roots, that is 5, and multiply this by itself, giving 25. From this subtract 21, giving 4. Take the root of this, 2, and subtract it from half of the roots, leaving 3, which represents one of the parts.

Today's algebra students learn this as the method of "completing the square" of the quadratic equation $x^2 - 10x + 21 = 0$.

The purely symbolic manipulation would be as follows:

$x^2 - 10x + 21 = 0$

$x^2 - 10x = -21$ (subtracting 21 from each side)

$x^2 - 10x + 25 = -21 + 25$ (adding 1/2 the coefficient of the middle term, squared, to each side)

$(x - 5)^2 = 4$ (noticing that the left side is a perfect square)

$(x - 5) = \pm 2$ (extracting the square roots of each side)

$x = 3$ and $x = 7$ (adding 5 to each side)

These days, there is nothing mysterious about al-Khwārizmī's method. Completing the square goes far back to classical Greek times when such questions were done by pure geometry and methods that had to be justified by axioms. We see no axioms in the *Almucabola*. Perhaps that is what makes it mysterious. The rules follow some kind of inclination that comes from balancing an equation and extracting roots, some kind of intrinsic logic that we cannot easily express by ninth-century logic.

What makes al-Khwārizmī's work different from Diophantus's? The writing in the *Almucabola* appears to be almost as rhetorical as that of the *Arithmetica*, except for a few minor abbreviation improvements, a notion and symbol for zero, and the Indian numerals. No new symbols were introduced. Rather, the *Almucabola* gave a list of problems organized according to various species of forms. Al-Khwārizmī says in the beginning, on his first page:

I discovered that the numbers of restoration and opposition are composed of these three kinds; namely roots, squares and numbers . . .

Of these three forms, then, two may be equal to each other, as for example:

Squares equal to roots
Squares equal to numbers, and
Roots equal to numbers.

By root, he means the unknown, what we would call x. By square, he means the square of the unknown, what we would label as x^2. Hence, by our symbolic notation, his example translates to:

$$ax^2 = bx,$$

$$ax^2 = c, \text{and}$$

$$bx = n, \text{where } a, b, \text{ and } c \text{ are positive.}$$

Al-Khwārizmī had given us methods for solving general linear and quadratic equations by reducing any one of them to one of three different forms. His methods simply involved collecting terms of the same species to one side of the equation by adding the same quantity to both sides. That is how algebra is done: complete and balance. (Equations in Al-Khwārizmī's time did not have sides—left and right—separated by some kind of equal sign as ours do, but the idea of completing and balancing as if there were sides would have made sense.) The symbols of algebra simply offer an easier way of carrying out the collection process; the use of symbols in mathematics does not make it algebra any more than words do. Once this is understood, the whole enterprise becomes one of seeing an infinite class of problems from the point of view of a finite form.

Such breakaway inspirations led to surprises.

Al-Khwārizmī wrote:

> When I considered what people generally want in calculating, I found that it always is a number.... What must be the amount of two squares which, when summed up and added to ten times the root of one of them, make up a sum of forty-eight dirhems?

Here, we would write $2x^2 + 10x = 48$.

You must at first reduce the two squares to one; and you know that one square of the two is the moiety of both. Then reduce every thing mentioned in the statement to its half, and it will be the same as if the question had been, a square and five roots of the same are equal to twenty-four dirhems; or what must be the amount of a square which, when added to five times its root, is equal to twenty-four dirhems?

Now halve the number of roots; the moiety is two and a half.

We would write this as $\frac{5}{2}$.

Multiply that by itself; the product is six and a quarter. Add this to twenty-four; the sum is thirty dirhems and a quarter.

We would write this as $6\frac{1}{4} + 24$.

Take the root of this; it is five and a half. Subtract from this the moiety of the number of the roots, that is two and a half; the remainder is three. This is the root of the square and the square itself is nine.[7]

We would write this as $5\frac{1}{2} - 2\frac{1}{2} = 3$.

"Take the root of this," al-Khwārizmī wrote. *The root,* singular! For him there was only one square root of $30\frac{1}{4}$—namely, $5\frac{1}{2}$. This led him to only one solution, the positive root $x = 3$. We would have done this slightly differently. We would use what high school algebra students call "completing the square," as we did at the beginning of this chapter:

Take $2x^2 + 10x = 48$.

Divide everything by 2 to get $x^2 + 5x = 24$.

Add $\left(\frac{5}{2}\right)^2$ to both sides to get $x^2 + 5x + \left(\frac{5}{2}\right)^2 = 24 + \left(\frac{5}{2}\right)^2$.

Simplify the preceding to get $\left(x + \frac{5}{2}\right)^2 = \frac{121}{4}$.

Take the square root of both sides to get $x + \frac{5}{2} = \pm\frac{11}{2}$.

And finally, subtract $\frac{5}{2}$ from both sides to get $x = 3$ and $x = -8$.

Whoa! What is that -8 doing here?

An understanding of al-Khwārizmī's solution was limited to the mechanics of algebra, because the mathematical strings connecting algebra and geometry were not yet understood.

Imagine al-Khwārizmī's solution drawn through the geometry of the quadratic equation. If he knew and used that geometry, he would have seen $x^2 + 5x = 24$ as a graph cutting through height 0 at two values of x—namely, 3 and −8 (see figure 11.1).

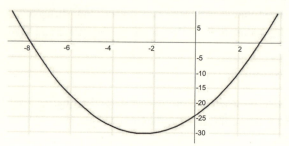

FIGURE 11.1 Graph of $x^2 + 5x = 24$.

Or, had he the symbolic algebraic tools to factor $x^2 + 5x = 24$, he would have noticed that his equation was the same as $(x-3)(x+8) = 0$, which has two solutions, $x = 3$ and $x = -8$.

Al-Khwārizmī knew about negative numbers from the *Brahmasphutmasiddhanta*, and even knew that there are two roots to any equation of the form $ax^2 + b = cx$. Brahmagupta defined zero as a number, the number one gets from subtracting a number from itself. In that way, he could list his arithmetical rules:

A debt minus zero equals a debt.
A fortune minus zero equals a fortune.
Zero minus zero equals a zero.
Zero minus a debt equals a fortune.
Zero minus a fortune equals a debt.
Zero times a debt or fortune equals zero.
Zero times zero equals zero.

By his language, we must conclude that he was thinking of negative numbers as numbers. His arithmetically logical rules talk of "fortunes" and "debts," and hence about positive numbers and a hint at the idea of negative numbers. He knew that in some cases a quadratic equation would have two roots, and that the condition of the application leading to the equation would exclude one. Even as late as Fibonacci's time, the negative root was suspect.

Until recently, it was assumed that Brahmagupta was the first mathematician to use negative numbers in any modern sense. But in China, negative numbers had been used since the beginning of the first millennium. They appear in *The Nine Chapters on The Mathematical Art* (see chapter 3). So the Chinese had them four hundred years before the Indians.

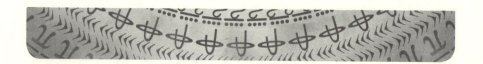

Chapter 12

Symbol Infancy

Algebra was not always called algebra. In the mid-fifteenth century some Italian and Latin writers called it *Regula rei e census* (*Ruling Out of the Thing and Product*). Mathematicians prefer short names for their fields—arithmetic, geometry, calculus, analysis, number theory, logic, and so on.

François Viète first called it the "analytical art." John Wallis gave it the English name "specious arithmetic." Most likely, his word for it came from the Greek word εἶδος, which meant "species," as well as the particular, *special* power of the unknown. The word "specious" was used to suggest that the species—monads, squares, cubes, and so on—generally represented all known and unknown quantities. In fifteenth-century English, the word "specious" meant pleasing to the eye in form, yet deceptive. The word was still in use with that meaning in the eighteenth century and could be found in Samuel Johnson's *Dictionary of the English Language* printed in 1785.[1] Newton called algebra "universal arithmetic," presumably because it embodied all the universal laws of arithmetic to be used on general equations. Petrus Ramus thought the Arabic name "algebra" was a vulgar name for "an art of [such] admirable subtlety." For Descartes, the Arabic name was "barbarous."[2]

At its surface, algebra seems to be the art of manipulating symbols according to some rules for doing so. But then, what should we make of a question that asks for the number that when added to its fifth gives twenty-one? Its verbal solution gives the answer as seventeen and one half. That particular question was asked in the Ahmes papyrus. The solution, as it was written on that papyrus sometime before

1550 BC, would not be recognizable to the student of algebra today, who would simply write $x + \frac{1}{5}x = 21$ and effortlessly manipulate the symbols by our modern rules of the art to solve for x, and get $x = 17\frac{1}{2}$. That student would probably be unaware of thoughts enabling generalization and unification and deeper levels of understanding, all coming directly from the notation itself.

The twentieth-century mathematician and science fiction author Eric Temple Bell once remarked—with little evidence—that in the mid-seventeenth century, mathematicians were able to introduce negative and rational exponents because symbolic manipulation liberated their thinking from the wilderness of words. In his classic popular history book, he wrote, "The more honor, then to the ancients who persevered through jungles of words to attain what the moderns reach with a few almost mechanical strokes of the pen."[3]

Evidence of this comes when we notice that for many centuries those mathematicians who worked their algebra rhetorically were not seeing what we now see. With the new symbolic representation of powers of the unknown $x, x^2, x^3, \cdots$, came hints of how the products of the powers of x were governed by the addition of the exponents. Bell wrote in his book *The Development of Mathematics*, "an incredible mass of confusing terminology and inefficient rules was swept into the past, and with it, equal or greater mass of torturous thinking."[4]

Students today are quite surprised to learn that before our relatively modern notion of a negative number, and before having a symbolic way of representing a negative number, the equation $ax^2 + bx + c$ was regarded as completely different from the equation $ax^2 = bx + c$. And each of these was still different from $ax^2 + bx + c = 0$. This may seem odd because, to us, the solution of one must be the solution to the others; however, before the sixteenth century, the arbitrary rational constant terms a, b, and c had to be positive.

For the most part, by our definition of symbol, Babylonian mathematics had no symbols other than their smart numerals and pictograms. Even Diophantus, as we have seen, did not make use of the full power of symbolic manipulation. Yet the art of algebra was practiced, for it was not—and never has been—limited to the mere manipulation of symbols. Algebra—by a rhetorical jungle of words or by symbols themselves—is the art of understanding relationships, of which "equality" is just one kind.

Every generation produces astounding people who fill awaiting niches that either forward human existence or make the world a better place for everyone—Johannes Gutenberg, Galileo Galilei, Leonardo da Vinci, Martin Luther King Jr., and Nelson Mandela, to name a few. I have no doubt that great mathematicians such as Newton and Leibniz would have been able to do fantastic mathematics in a world without a single symbol. Also, I have no doubt that in such a world their work would have presented serious struggles, and possibly almost insurmountable obstacles—*almost*, because almost nothing is insurmountable for humans. So what brought on the flood of symbols and algebra notation in the second half of the sixteenth century, a gush powerful enough to completely change the way mathematics and science has been done since the seventeenth century?

Imagine where we would be today if algebra were still entirely rhetorical, or even just abbreviated. Students of algebra fear those "word problems" that fill chapters of their elementary algebra texts. They are easier than those problems that students in the fifteenth century had to solve, for the modern student knows that all that has to be done is to translate the problem into symbolic notation, and let the rules of symbolic manipulation take it from there.

Abacus algebra began in Fibonacci's time and flourished from the mid-fourteenth century. It was a tradition of problem solving that came from the *abacus* schools and the *maestri d'abbaco* treatises that dealt with large numbers of arithmetic and algebraic problems backed by rules or geometric demonstrations. That tradition was partly responsible for the emergence of symbolic algebra that began in the first half of the sixteenth century.

The ideas of algebra brought on the symbols, not the other way around. Robert Recorde had written the words "is equal to" almost two hundred times in his book *Whetstone of Witte* (1557) before noticing that he could easily "avoid the tedious repetition" of those three words by designing the symbol ===== to represent them.

The initial incentive was the need to abbreviate, but once the equal symbol was in place, something else took over. The concise character of the symbol came with an unintended benefit: it enabled an unadorned picture in the brain that could facilitate comprehension.

Early historians had credited the Arab algebraist al-Qalasādi as being the first Arab to use letters of the Arabic alphabet to denote arithmetic operations. He was born in Bastah, a Moorish city in what is now northeast Spain, where he studied law and the Koran. Later, when the Castilians began their conquering push eastward, he moved south to Granada in Andalusia. In the early part of the fifteenth century, almost the whole of Spain and Portugal was Muslim and at constant war with the Castilian and Aragonian Christians. Al-Qalasādi wrote several books on arithmetic and one on algebra that had mathematical notation made from shortened Arabic words and letters. Such notation was indeed used in his treatise on algebra, *Al-Tabsira fi'lm al-hisab* (*Clarification of the Science of Arithmetic*).

His notation was clearly an attempt at symbolizing algebra through abbreviations, a first approximation to what we would consider true symbols; however, we should also be aware that they were already used by North African Muslim mathematicians for at least a century: he was not the originator. A century before al-Qalasādi, however, the Maghrebian mathematicians Ibn al-Banna and Ibn al-Yasamin also had schemes for a kind of abbreviated alphabet notation, and surely such alphabetical symbols had been used in the East far earlier than the thirteenth century.[5]

Small things happen now and again to move the progress of human intelligence and to benefit the world. Think of the language tools that have emerged alongside the modern programmable computer revolution that has been with us since the 1940s, and how quickly they changed to give us the modern laptop and café. Could we have all those laptop apps relying solely on low-level machine or assembly languages without a general-purpose computer programming language? Sure! But pity the people who have to do the work, and imagine how long it would take. Tedium and complexity would call for more gifted people than we have schooled.

And so it was with the symbol childhood of the sixteenth century. Could we have all the modern mathematics and physics we now have without the symbolic language that was developed in the late sixteenth century? Sure! But what would it have taken to have that same achievement? At least a huge dedicated math populace.

Italy wasted no time in cultivating the seeds of algebra that drifted to Europe after the Arabs brought that art to Spain. Unfortunately, except for the works of Fibonacci, almost nothing is known about European works of algebra before 1300. The earliest works of that period were those of Fibonacci, Paolo de l'Abacco, and Belmondo de Padua. By the end of the fifteenth century, algebra went no further than quadratic equations with just one unknown, a level close to the syllabus of any present-day high school course. Back then, the art was still being performed rhetorically; there were still no signs or symbols for the unknown or operations, and quadratics had only the positive roots.

In 1505, Scipione de Floriano de Geri del Ferro (ca. 1465–1526), more commonly known as Scipio del Ferro, solved a specific case of a compound cubic equation, the case $x^3 + ax = b$, where a and b are positive numbers. At that time, negative numbers were still under suspicion and therefore not used. The same for zero, which was still regarded suspiciously, though four centuries had passed since its introduction to Europe. So the equations $x^3 + ax = b$ and $x^3 = ax + b$ were considered different. In our symbolic algebra, which uses negative numbers and zero, those two cases are not too far from the general case $x^3 + ax^2 + bx + c = 0$.[6]

Del Ferro's solution—like much of the mathematics of the time—was not done with literal coefficients, but rather with strategic choices of convenient numbers. Working out the formula for the particular cubic polynomial $x^3 = 9x + 28$ and finding the solutions to be $x = 4$ and $x = -2 \pm i\sqrt{3}$ gave confidence in finding a general method. The mind develops its own ways for dealing with abstractions through repeated examples, yet, mysteriously, it is also capable of generalizing from a single example. Without symbols, it would certainly not be easy. Along the way, it would be clear that the coefficients 9 and 28 might be convenient for computational purposes, although not confining enough to hamper any general procedures. That was how algebra was done in del Ferro's time.

That wonderful substitution, $y = x - \frac{a}{3}$, the one that reduces the general cubic equation to one without a quadratic term, was a general procedure known to del Ferro, though it was performed geometrically and only on specific example cubic polynomials, so a would have been the coefficient of a particular quadratic term, a specific number. Ideas for substitutions are now so elemental to the modern algebra choreography of reducing one problem to a simpler one that we must marvel at its formidable brilliance and wonder how such a work of genius could possibly be done without the use of symbols. Not an idea that easily lends itself to rhetorical expression, it surely is one of the grand benefits of the symbolic approach. And so del Ferro was able to construct the general solution to the cubic equation $x^3 = ax + b$ as:

$$x = \sqrt[3]{\frac{b}{2} + \sqrt{\frac{b^2}{4} - \frac{a^3}{27}}} + \sqrt[3]{\frac{b}{2} - \sqrt{\frac{b^2}{4} - \frac{a^3}{27}}}.$$

If this looks a bit menacing, just imagine how fearfully tormenting it must have been in del Ferro's time, or even a hundred years after, without the symbols of our modern notation. Applied to $x^3 = 9x + 28$, where $a = 9$ and $b = 28$, the entire right side collapses to give $x = 4$. Del Ferro did not know that there must be three solutions, but let's leave that for later, when we take this up again in chapter 14.

By 1545, the Italians—in particular, Gerolamo Cardano, his student Lodovico Ferrari, and his rival Niccolò Fontana Tartaglia—had solved general cases of cubic and quartic equations. Cardano's *Ars Magna*, short for the more formal title *Artis Magnae, Sive de Regulis Algebraicis Liber Unus* (*The Great Art, or the Rules of Algebra, Book Number One*), was published in that year. It contained everything that was known about cubic and quartic equations up to that time, including (for the first time in print) both real and complex numbers as roots (called "true" and "fictitious") to those cubics. It gave the rules geometrically, and acknowledged that Niccolò Tartaglia communicated the rules—not the proof—for solving cubics to Cardano, but that Tartaglia learned the rules from del Ferro. All algebra of that time was still mostly rhetorical algebra with a few symbols creeping in, such as $\text{R}\!\!\!\!x$ for root. Presumably $\text{R}\!\!\!\!x$ was a shorthand for the Latin word for "root," *radix*; square roots of negative numbers would have been written as $\text{R}\!\!\!\!x.\tilde{m}.$, so $\text{R}\!\!\!\!x.\tilde{m}.15$ would have meant $\sqrt{-15}$. Del Ferro communicated his ideas to a small circle of friends, students and colleagues, but never published. Therefore, we have no direct evidence of his work, just that which comes from Cardano, himself, who traveled from Milan to Bologna to meet del Ferro.

Though the story behind the main achievement of his great work is recognized as one of the most ferocious feuds in the history of mathematics, Cardano did give credit to others, including his old friend Niccolò Tartaglia.[7] Referring to the solution of $x^3 = ax + b$, Cardano gave those acknowledgments in the first chapter of his *Ars Magna*:

> In our own days Scipione del Ferro of Bologna has solved the case of the cube and first power equal to a constant, a very elegant and admirable accomplishment. Since this art surpasses all human subtlety and the perspicuity of mortal talent and is a truly celestial gift and a very clear test of the capacity of men's minds, whoever applies himself to it will believe that there is nothing that he cannot understand. In emulation of him, my friend Niccolò Tartaglia of Brescia wanting not to be undone, solved the same case when he got into a contest with his [Scipione's] pupil, Antonio Maria Fior, and, moved by my many entreaties, gave it to me.[8]

The feud was not over acknowledgment; rather, it was over his pledge of secrecy to Tartaglia, and a promise to not publish it.[9] Cardano gave his version of the solution, a purely geometric one, because he did not have the symbolic tools necessary to do the hard work. For example, we can show that $(a+b)^3 = a^3 + 3a^2b + 3ab^2 + b^3$ by multiplying three copies of $(a+b)$, using a few rules of algebra. Cardano had no such advantage. He had to diagrammatically slice a geometric cube to get the same result.

His rhetorical representation of equations had one great drawback. It required long lists of equation types, in part because he would not use zero. He was suspicious of the rules of signs for the multiplication, arguing twenty-five years after the publication of *Ars Magna* that minus times minus makes plus has the same truth as telling that plus times plus makes minus.[10] In reflection he writes, "And therefore lies open the error commonly asserted that minus times minus produces plus, lest indeed it be more correct that minus times minus produces plus than plus times plus would produce minus."[11] So it is hard to understand how he could have such suspicion and still have argued for negative solutions to linear problems and accept square roots of negative numbers.

The *Ars Magna* was a real breakthrough for algebra. Though it didn't have the symbols that were soon to be invented, and therefore was excessively strenuous to write and cumbersome to read, it gave mathematicians the chance to see a real need for a better source of symbols to make comprehension easier and stronger. Geometry in those days tended to be a subject of visual logic; one had to see a drawing, at least in the mind's eye, in order to approve of the inherent logic. It had been that way ever since the time of the Pythagoreans, Euclid, Apollonius, and Archimedes, when many problems were solved by geometric means—lines, squares, and cubes imagined in space—with geometric proofs, partly because there were few other means available.

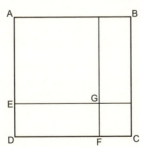

FIGURE 12.1 Viewing $x^2 - 2ax + a^2 \equiv (x - a)^2$ geometrically.

To give an example, here is a purely geometric reasoning that would show the algebraic identity $x^2 - 2ax + a^2 \equiv (x - a)^2$. First consider the diagram of the square ABCD (figure 12.1). Let x represent the length of side AD and a the length of segment ED. Then x^2 is the area of the square with side AB, ax the area of the rectangle with sides ED and DC, and a^2 the area of the small square with sides GF and FC. By subtracting the area represented by ax twice from the larger area represented by x^2, we would *almost* have the area of the square with side AE. Almost, because it would be deficient by the area of the small square of side GF (which we subtracted twice, when we should have subtracted only once). So we need to add back the area of the small square with side GF. Algebraically, what we did was to subtract $2ax$ from x^2 and add back a^2 to get $(x - a)^2$. Hence, $x^2 - 2ax + a^2 \equiv (x - a)^2$.

Cardano did not have our distinct advantage of working with literal numbers as the coefficients. His treatment would have meant a specific assignment for the size of a in proving the preceding identity. So he would deal with an equation of the form $ax^2 = bx + c$ by choosing a, b, and c to be positive numbers appropriate to his example—say, 1, 10, and 144. In his first example (*Ars Magna*, chapter V), he tells us that $x = 18$ is *the* solution (ignoring the negative solution $x = -8$) to $x^2 = 10x + 144$.

If Cardano had the symbolism necessary to perform the algebra directly, he might have done what we would do. Balancing equations by rules, he would have taken $(x - a)^2$ to be $(x - a)(x - a)$ and used the distributive law from the right to get $(x - a)x - (x - a)a$. Then he would have used the commutative law to get $x(x - a) - a(x - a)$, and once again use the distributive law to get $x^2 + (-ax) + (-ax) + (-a)(-a)$. But here he would have been stuck, for he would have to agree that $(-a)$ times $(-a)$ equals $+a^2$. He would instead argue that the $+a^2$ that comes in the end is not a result of the fact that $(-a)$ times $(-a)$ equals $+a^2$, but rather because it is an area that had to be replaced after it was subtracted twice when it should have been subtracted only once. He might even quote Euclid's *Elements*, book II, proposition 7, for the security in knowing that one can perform the geometry of subtracting and adding squares and rectangles the way he did to get the desired result. Even if he had the proper symbols at hand, he would have had to have the algebraic rules to justify any manipulation he wanted to perform; such rules were not fully in place, yet.

When it came to a detailed proof of solutions to polynomials higher than cubic, geometry was no longer much help. "It would be foolish to go beyond this point," he wrote, "Nature does not permit it."[12] So the *Ars Magna* was a struggle to write and a struggle to comprehend, especially when it came to solving higher degree equations. Cardano himself told us that it was very difficult.[13]

The most remarkable part of Cardano's treatise is that, though rhetorical and cumbersome, it clearly recognized the value of imaginary and complex solutions, which had been completely avoided by earlier writers. Even though he would not multiply two negative numbers to get a positive number, he would have no qualms about multiplying and dividing two square roots.

Algebra was dragged along by the early language and notation that hindered a certain way of thinking symbolically. It is, therefore, hard to determine the precise moment when our way of algebraic thinking actually began. Imagine trying to think with unknown quantities as well as operations on quantities all expressed by their full-length names. Humans hate tiresome repetition; when the repetition goes too far they search for simplification.

I recall one summer when I was just fifteen and worked (illegally) for a week at my uncle's silkscreen workshop. Like Charlie Chaplin in *Modern Times*, all day long, from 8:30 to 4:30, I stood in one place like a simple-circuited robot, moving ink-wet posters from a silk screen to a drying rack, one after another: *extend both hands, grab corners, lift, turn, slide into rack, release corners, turn back, repeat.* The grand highlights of the boring hours happened when a rack was full and another had to be rolled into place, a break in the robot's program that gave a chance to walk a step or two. I counted the minutes along with the wet posters being moved, one-by-one. To make it even more painful, a great big Seth Thomas clock that might have been originally designed for a train terminal was mounted high on the wall directly in front of the drying rack, the slowest moving clock in the world.

Every night, I would invent and sketch a new mechanical contraption that might someday lighten daytime monotony for human conveyor belts. If only I had prudently patented my ideas...

Humans have always made tools or invented machines to do repetitive work; they conceive abstract solutions for recurring questions; they concoct shorthand for tiresome verbosity. And so, after writing thousands of repeated arithmetical words, mathematicians caught on to the idea of substituting initials of words for the words themselves.

In the words of Ernst Mach:

> Strange as it may sound, the power of mathematics rests upon its evasion of all unnecessary thought and on its wonderful saving of mental operation. Even those arrangement-signs which we call numbers are a system of marvelous simplicity and economy. When we employ the multiplication-table in multiplying numbers of several places, and so use the results of old operations of counting instead of performing the whole of each operation anew; when we consult our table of logarithms, replacing and saving thus new calculations by old ones already performed; when we employ determinants instead of always beginning afresh the solution of a system of equations; when we resolve new integral expressions into familiar old integrals; we see in this simply a feeble reflexion of the intellectual activity of a Lagrange or a Cauchy, who, with the keen discernment of a great military commander, substituted new operations for whole hosts of old ones. No one will dispute me when I say that the most elementary as well as the highest mathematics are economically-ordered experiences of counting, put in forms ready for use.[14]

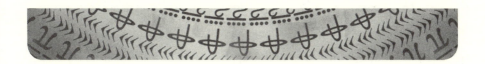

Chapter 13

The Timid Symbol

In Nuremberg, Germany, a year before Cardano's *Ars Magna* appeared in print, people were studying Michael Stifel's *Arithmetica Integra*, a treatise on arithmetic and algebra. Stifel included several symbols that were already in use, such as +, −, and √, which he actually called "plus," "minus," and "radix"; still, there was no sign for "equals."

Symbols were beginning to appear in European manuscripts on algebra in two different styles: one from the Italians, the other from the Germans. The Italians used the word cosa ("what," or "thing") when referring to the unknown root of an equation. And since algebra was after all the art of finding such *cosa*, the northern Europeans began calling algebra the Cossic Art.

The oldest notation for radicals (square roots, cube roots, and so on) dates back to about 1480, when a dot placed before the radicand (the quantity to be square-rooted or cube-rooted) was used to signify a square root: two dots for the fourth root, and three dots for the cube root. By 1524, the dot evolved into a blackened point with a tail bent upward to the right. It looked very much like a musical note, something like $\int$. But this new symbol wisely didn't carry along the old idea of repeating itself to get cube and fourth roots. Instead, it had another symbol attached that indicated the rank of the root. So, $\int 2_{\mathcal{Z}}$ would indicate square root of two, and $\int 2c^e$ would indicate the cube root of two. To indicate the fourth root, an extra dot was used. Accordingly, $\bullet\int 2_{\mathcal{Z}}$ would indicate the fourth root of two.[1]

Algebra at that time was concerned with solving cubic and higher degree polynomials. Solutions often involved binomials that, in our notation, would appear as $\sqrt{x} + \sqrt{b}$. To indicate that the root is to apply to the entire binomial, an elongated L was used along with *cs*, an abbreviation for the Latin *communis* ("common"). The base of the L would extend under the binomial. For instance, $\int cs|3+\int 2\tilde{g}$ would indicate $\sqrt{3 + \sqrt{2}}$, the plus sign being already occasionally used in German manuscripts.

Michael Stifel's edition of Christoff Rudolff's *Die Coss* (1525) incorporated the symbol ✔ (a close match to our $\sqrt{\ }$) for square root, along with 〰 for cube root and, peculiarly, 〜 for fourth root.[2] So, 〰 64 = 4 and 〜 16 = 2. Peculiar, because these symbols must have contributed to misunderstandings suggesting that ✔✔✔ 64 equals 4 and that ✔✔ 16 equals 2. The second is true. The first is not. Clearly, Rudolff did not mean to connect any significance to the similarity between 〰 and ✔✔✔, and yet historians have mistaken his intent by writing his symbols for cube and fourth root as $\sqrt{\ }\sqrt{\ }\sqrt{\ }$ and $\sqrt{\ }\sqrt{\ }$. This confusion shows how difficult it is to design good symbols.[3]

Rudolff's symbol for square root had another disadvantage. How could his readers distinguish between $\sqrt{\sqrt{512} + 16}$ and $\sqrt[4]{512} + \sqrt{16}$? They had to watch out for dots. To indicate that he meant $\sqrt{\sqrt{512} + 16}$, he would write ✔.✔ 512 + 16, the dot indicating a grouping that extends over the next term.[4]

Perhaps *Die Coss* was the first complete algebra text written in the German language, a reference work for all that was known about algebra up to the early sixteenth century, and a fabulous resource for future textbook writers. At the time of its printing, most symbols were still mere abbreviations of words without any standardized agreement. Although + and – were used on occasion, so were *p* and *m*. And ✔ could be marked as R, or ℞, or *res(x)*, which was Latin for "thing," meaning the unknown thing that was to be found.[5] For many years after the publication of *Die Coss*, historians took the symbol ✔ as a rapid writing of the letter "r." Euler thought so.[6] As we have seen, however, the symbol may have evolved from the German manuscripts where the dot evolved to become the symbol $\int$, a kind of dot with a tail, perhaps the trail a stylus would make after marking the dot quickly. The original symbol

✔ found in Michael Stifel's edition of Rudolff's *Die Coss* has no horizontal bar that gives the symbol an "r"-like image.

There is an unpublished manuscript dated 1344, the *Aliabraa arbibra*, attributed to someone called Maestro Dardi di Pisa. Almost nothing is known about the author beyond the fact that Dardi was an abacist who played an important role in medieval mathematics by introducing abbreviated notation. Even his name is obliterated from the manuscript, which is believed to be the earliest manuscript written in the Italian vernacular that exclusively treats algebra.[7] The *Aliabraa arbibra* uses abbreviations such as ℞ for radix, m̂ (the first letter of *meno*, the Italian word for "less"), *c* for the unknown (the first letter of *cosa*), and *ce ce* for fourth power instead of *censo di censo* (square of the square). Operations are not symbolized; however, there are some curious diagrams that must have been used to teach multiplication.[8] Dardi's calculations forced him to consider what to do about nested square roots such as $\sqrt{x + \sqrt{12}}$, which he would have written as "℞ de zonto censo co ℞ de 12."

After a bit of tinkering by Stifel in 1553, a modified version of Rudolff's symbol for square root was the one that stuck. So by 1570, the German symbol $\sqrt{}$ found its way through Europe, west to France and England, and south to Spain and Italy.

Many authors going back to Pacioli used the notation ℞, which soon took on the more cursive design ℞, which usually meant the root of the polynomial. Nicolas Chuquet used it (superscripted with a 2) in his algebra text *Triparty en la Science des Nombres* (*A Three-part Book on the Science of Numbers*), but not as the root of the polynomial; he wanted it to symbolize the square root.[9] We don't know exactly when the *Triparty* was written. The year of the writing would have been around 1484, but the manuscript was found in the 1870s, and since almost four hundred years went by before it was printed, it could not have had much influence on the historical development of notation, except for the fact that Chuquet's student Estienne de la Roche covertly copied the work and published it as his own, getting the credit of having the first French algebra book.

For Chuquet, a number was a first root. If he were to write ℞¹9, he would mean the number 9. If he were to write ℞²9, he would mean 3. In 1484, Chuquet wrote:

The root of a number is a number which, multiplied by itself one or several times according to the demands and nature of the root, produces precisely the number of which it is the root. Or otherwise the root of a number is such that, written and set down two or more times one under the other or one beside the other, and then the first multiplied by the second, and what comes by the third if there is a third, and again the fourth, and again by the other if there are others, the last product shall be equal to the number, or shall produce the number whose root it is. And one should know that there are infinitely many kinds of roots, for some are second roots, others third roots, others fourth roots, others fifth, and thus continuing without end.[10]

Chuquet's notation R is confusing when reading other works of the period because unfortunately its ancestor R had a double meaning. Sometimes, R was used to mean what we would label as x^2, and at other times, it would mean what we would call x. The confusion comes from the Latin words *latus* ("side") and *radix* ("root") used by writers working from Greek sources where all algebra was veiled in geometry. If we consider a geometric square, does R signify its side x or the root of its side, $\sqrt{x}$? To make matters worse, the notation would sometimes refer to the root of the polynomial, which for us would be the possible values of the unknown. The only way to tell was by context, so readers had to be attentive. The problem stemmed from Pacioli, who used R for both roots and powers. In his *Summa* (1494), he wrote[11]

$$\mathrm{R}.p^a = x^0,$$
$$\mathrm{R}.2^a = x, \quad \mathrm{R}.3^a = x^2, \quad \mathrm{R}.4^a = x^3$$

and

$$\mathrm{R}.2 = \sqrt{2}, \quad \mathrm{R}.3 = \sqrt{3}, \quad \mathrm{R}.4 = \sqrt{4}.$$

Such awkward notation leads to problems. In the second line preceding, the numbers are always one unit higher than the powers they represent. We want x^n to mean multiply x by itself n times. That is what it should mean so that we get our wonderful little multiplication formula for exponents, $x^m \cdot x^n = x^{m+n}$. Pacioli's notation has that unnecessary minus one as exponent:

$$\mathrm{R}.m^a \cdot \mathrm{R}.n^a = x^{m-1} \cdot x^{n-1} = x^{(m-1)+(n-1)} = x^{m+n-2} = \mathrm{R}.(m+n-1)^a.$$

We can ignore the inelegance. The formula works, and $R\kern-0.6em\raise0.9ex\hbox{.}n^a$ surely does represent a power of x, just not n copies. But inelegance is not the only concern. There does not seem to be a symbolic way of raising a power to a power—that is, there does not seem to be a way to write $(x^m)^n$ without resorting back to rhetoric. We would not have our beautiful formula $(x^m)^n = x^{m \cdot n}$. In other words, the awkwardness of the notation is an obstruction to more advanced symbolic representation.

An obstruction, yes. But with some benefit. It unleashed the terms of algebraic expressions from their ancient geometric metaphors. The square and cube of a number had been mentally linked to the geometric square and cube since Babylonian times. Euclid used the word δύναμις ("power," the word we use when talking about exponents) to talk about the square. He would say that two magnitudes are "commensurable in power" to signify that the lengths of two distinct lines could be measured by the same yardstick (in Euclid's case, the same "pygon stick"). It is not possible to measure the diagonal of a square and the side of a square with the same yardstick. So "power" for Euclid was a purely geometrical word, and not the number formed by multiplying a number by itself.

Chuquet's numerical notation R^1, R^2, R^3, R^4, ... went beyond the possibilities for geometric metaphor of dimension. One immediate advantage is that we wonder about how such a notion would extend to negative numbers. Indeed, Chuquet did wonder. He would write $R2^{\overline{1m}}$ to signify $2x^{-1}$, the exponent form of $\frac{2}{x}$. Such notation forces x^0 to be equal to 1, a relation that Chuquet knew and used.

In a roundabout way, the *Triparty* should have contributed to an early development of modern exponential notation, supercharging progress in algebra.[12] But it did not. It was a hundred and fifty years ahead of its time; however, because it was not printed and distributed, Chuquet and his *Triparty* were unknown to mathematicians.[13] And so the whole idea of negative exponents had to wait for its publication some hundred and seventy years later, when John Wallis used negative exponents in his *Mathesis Universalis*.[14]

Stifel published an edition of *Die Coss* in 1553, in which our equation $x^2 - 3x + 2$ appears as $1Zm.3Rp.2$. In such notation, the unknown is Z, and it stands for x^2. It came from the word *zensus*, an alternative old German spelling of the Latin *census*, meaning "number of."[15] The R represents the square root of Z. In our notation, that would mean $\sqrt{x^2}$, which, of course, is always positive. This must have hindered all thoughts of negative roots that a better notation would have exposed, and eliminated at the gate any roots that may end up being imaginary.

By 1575, that same expression ($1Zm.3Rp.2$) had become $1Q - 3N + 2$ and then later $1AA - 3A + 2$. The advantage to this last notation is that it clearly exhibits the relationship between the first two terms—that is, one notices at a glance that there is almost nothing different between the first and second terms other than their powers. By writing the polynomial as $1AA - 3A + 2$, we see the relationship between the first two terms, but we don't "see" that relationship when viewing that same polynomial as $1Q - 3N + 2$. Our modern notation for this polynomial is $x^2 - 3x + 2$.

Chapter 14

Hierarchies of Dignity

"I feel obliged to speak of the supremacy, among all the mathematical disciplines, of the subject that is nowadays called algebra by the common people," Rafael Bombelli wrote in 1572.[1] Bombelli was an engineer whose work involved something to do with reclaiming marshlands and building bridges. His *L'Algebra* was published in 1579, but he began working on it twenty years earlier, when he had a break from his work at draining the Val di Chiana marshes in central Tuscany. In *L'Algebra*, we meet a new kind of notation for the unknown and its powers. Our modern notation for the polynomial $x^2 - 3x + 2$, for instance, evolved through several intermediate stages from Cardano's time to Descartes's. By our standards, the intervals between stages were many and long.

Twenty-seven years before Bombelli published *L'Algebra*, when Cardano was still writing his *Ars Magna*, equality was not used with any subordination in mind. The Latin *aequalis* was used liberally to tell us when two expressions were the same, and that they could be swapped without any loss. The two sides of *aequalis* had equal rank.

A few years later, in his *Whetstone of Witte*, Robert Recorde introduced his Gemini markings to northern Europe. It was an elongated form of our "equals" sign, a symbol that shows up less often than "+" and "–", yet more often than any other in the vast collection of the whole world's mathematical writing.

> And to avoide the tediouse repetition of these wordes: is equalle to: I will sette as I doe often in woorke use, a paire of paralleles, Gemowe

lines of one lenghte, thus: ====, because noe. 2. thynges, can be moare equalle.[2]

That Gemini, that twin, that supreme symbol of equation, is an inspired gift to our modern symbol collection. It was brilliantly designed (possibly by the Italians before Recorde first used it) to keep the notion in mind that there are two things that are meant to be exactly the same—a simple invention to help the reasoning process.

Bombelli, who surely knew of Recorde's *Whetstone of Witte* was still writing the word *fa* ("makes") or *eguali* ("equal") when he wanted to tell us that one expression begat another. You will not find an equal sign in *L'Algebra*. Rarely do we find the words *è eguale a* between two expressions. We use the term "equals" when we wish to say that for some useful purpose two things are swappable and that it makes no difference which one you use for that practicality. We might say that four quarters equals one dollar. Quarters are made mostly from copper. Dollar bills are made mostly of wood pulp and cloth. In physical appearance, they are not the same. For some practical purposes, such as buying chewing gum at a convenience store, there is no difference. But pay the bill for two people at a fashionable restaurant with quarters, and you will be aware of a difference.

L'Algebra, written in Italian, used equality in a different sense than we do. Words such as *fa, faro, eguali,* and *eguale* are unidirectional. Saying *sommato 1 uia 1, farà 2* ("the sum of 1 and 1 makes 2") is not quite the same as saying "1 plus 1 equals 2," for the Italian suggests that "2" is subordinate to "1 plus 1." There is a conceptual difference that obstructs the notion of balance between the two sides of the equality $1 + 1 = 2$. The Latin *aequales* means "equals," a word that maintains unbiased duality between perfectly swappable entities, but Bombelli chose to use the unidirectional *fara*. What makes Recorde's notation so appropriate and advantageous is that it suggests the very duality mathematicians were looking for when they wanted two things to be freely exchangeable without any unintentional suggestion of subordination.[3]

In 1572, Bombelli would have written $x^2 - 3x + 2$ as 1.$\overset{2}{\smile}$m.3.$\overset{1}{\smile}$p.2, or sometimes as 1.$\overset{2}{\smile}$m.3.$\overset{1}{\smile}$p.2$\overset{0}{\smile}$, and sometimes, when performing an arithmetical operation between two polynomials such as multiplication, where he wanted columns to line up, would have written $\overset{2}{1}.\text{m}.3.\overset{1}{\text{p}}.2$. We see in figure 14.1 how he was able to square the polynomial $-4x^2 + 5x + 2$ to correctly get $16x^4 - 40x^3 + 9x^2 + 20x + 4$.

FIGURE 14.1 *L'Algebra*, cover and book II, page 217, *Centro di Ricerca Matematica Ennio De Giorgi.*

Examine how Bombelli performed this multiplication to see that he was using the symbols to perform the algebra. (See figure 14.1.) Below the first horizontal line, there are two columns, one marked *più* ("more," meaning "plus"), the other marked *meno* ("less," meaning "minus"). In the *più* column, he added the 2s to get 4, then those coefficients of *dignità* 1 to get 10, and then 10 again. Next, he multiplied the coefficients of *dignità* 1 to get 25 moved up to *dignità* 2. And then he multiplied the coefficients of *dignità* 2 to get 16 bumped up to *dignità* 4. That completed all operations in the *più* column. Similar multiplications and additions in the *meno* column completed the task. Adding up while not mixing *dignità* gave the correct answer as:

$$16\overset{4}{.}\text{p}.9\overset{2}{.}\text{p}.20\overset{1}{.}\text{p}.4.\text{m}.40\overset{3}{.}$$

What we call "exponents," he called *dignità*. The modern Italian *dignità* translates to the English "dignity," which may seem to be a strange word for what we would call "power," or "exponent." For him, the higher powers meant hierarchies of dignity.

He began book II with "Nomi delle dignità, e forma delle lore abbreviature" (Names of dignity, value, and form of abbreviations). Then he listed the *dignità*:[4]

Tanto ⌣

Potenza ⌣

Cubo ⌣

Potenza di potenza ⌣

Primo relato ⌣

⋮

Cubo di potenza di potenza ⌣

Personally, I prefer "dignities" to "exponents"; the word "exponent" comes from the Latin *ex-ponen*, which I interpret to mean "upward-placing," where "upward" suggests a hierarchy, or ranking of the powers. The word "dignity" in Old English means "rank"; Shakespeare used it to distinguish two things that are alike yet different in rank:

> But clay and clay differs in dignity,
> Whose dust is both alike.[5]

Bombelli, as we see, was not only inventing genuine symbols when depicting *dignità* as little cups holding numbers but also inventing words that were new to mathematics.

These clever authentic symbols gave algebra its independence from geometry. For the first 200 pages of his manuscript, Bombelli used the traditional way of writing polynomials, replacing the Z with a Q for *quadrato*, so $1Zm.3Rp.2$ became $1Q.m.3R.p.2$.

Ranking by numerical hierarchy made the rules of exponent multiplication more apparent. With the new symbols, it was easier to see that

$$1^{n} \text{ times } 1^{m} \text{ equals } 1^{n+m}.$$

But, then, why did Bombelli give us the following long list of redundant products of *dignità*?

It looks as if figure 14.2 represents a doodling account of sheep to put modern readers to sleep. The hard truth is that we tend to look at everything from the point of view of what we already know. We know that $x^n x^m = x^{n+m}$, because by definition x^n is nothing more than the product of n copies of x, and, therefore, all we have to do is count copies of x to convince ourselves that we could go on forever with any positive integer values of n and m. Such a general notion, however, would have been strangely foreign to readers in Bombelli's world.

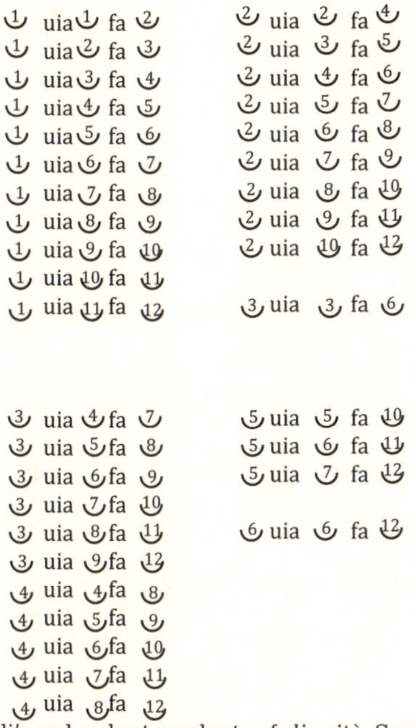

FIGURE 14.2. Bombelli's redundant products of *dignità*. Source: *L'Algebra*, book II, 205–206.

Bombelli was concerned with certain solutions to the cubic equation $x^3 = ax + b$, whenever a and b are positive numbers. One example would be the equation $x^3 = 9x + 28$, encountered in chapter 12. A solution comes from a simple guess that $x = 4$ works. But what about the two other solutions that would force us to do absurd things, such as taking a square root of a negative number?

The solutions to the general cubic of the form $x^3 = ax + b$ was given to us by del Ferro as:

$$x = \sqrt[3]{\frac{b}{2} + \sqrt{\frac{b^2}{4} - \frac{a^3}{27}}} + \sqrt[3]{\frac{b}{2} - \sqrt{\frac{b^2}{4} - \frac{a^3}{27}}}.$$

(See chapter 12.)

Apply del Ferro's formula to the equation $x^3 = 9x + 28$, where $a = 9$ and $b = 28$. It yields $x = 4$. Where are the other two solutions that we modern mathematicians know about? On the surface, del Ferro's formula gave us just one solution, even though the square roots inside the cube roots are not negative. We said $x = 4$, but if we actually substitute 9 for a and 28 for b, we find that $x = \sqrt[3]{27} + \sqrt[3]{1}$. Yes, $\sqrt[3]{27} = 3$ and $\sqrt[3]{1} = 1$. But if we want to actually find $\sqrt[3]{27}$, we would set $x = \sqrt[3]{27}$, and try to find x. That would mean finding a solution to the equation $x^3 = 27$, and by our modern notation we would know that $x^3 - 27 = 0$. The left side splits into the product $(x - 3)(x^2 + 3x + 9)$.

This last equation has three solutions, one coming from $x - 3 = 0$, and the other two coming from the quadratic equation $x^2 - 3x + 9 = 0$ When we add all solutions to $x = \sqrt[3]{27}$ and $x = \sqrt[3]{1}$, we get $x = 4$, and $x = -2 \pm \sqrt{-3}$, all three solutions to $x^3 = 9x + 28$.

Imagine Bombelli encountering these strange things that are square roots of negative numbers. What happens in his *L'Algebra* book II is fascinating, but if we explore this further we would be steering off our intended direction. For an easy diversion in that direction see chapter 7 of Barry Mazur's wonderful ("outstanding"—if I could really get away with my sincere but suspiciously nepotistic superlatives) *Imagining Numbers*.[6] For now, let us be satisfied with knowing that Bombelli encountered and accepted imaginary numbers, and even created an abbreviated notation for them.

A contraction of the *più radice di meno* ("more of the minus root") became *più di meno* and further abbreviated to *p.dm.* So $\sqrt{-2}$ would be written as *p.dm.2*, and $\sqrt{-1}$ would simply be *p.dm.* It would be a long time before the symbol i arrives to represent $\sqrt{-1}$, but *p.dm* was quite an advance because arithmetic errors are less likely to happen with proper notation designed to avoid them. The mistakes of writing $\sqrt{-2}\sqrt{-3} = \sqrt{6}$ and $\sqrt{-1}\sqrt{-4} = 2$ that Euler would make two hundred years later would have been avoided with notation that would mark the first product as $(i\sqrt{2})(i\sqrt{3})$ to get $-\sqrt{6}$, and the second product as $i(i\sqrt{4})$ to get -2. Of course,

Euler must be excused, as he was blind at the time these mistakes happened in the printing of his *Elements of Algebra*.[7] Although Euler was the one who introduced us to the symbol i for the imaginary number $\sqrt{-1}$, it didn't appear again until Gauss used it in 1867.[8]

Simon Stevin would have written the polynomial $x^2 - 3x + 9$ as 1②-3①+9⓪. Somehow, the *cosa*, the "root," the "thing"—whatever the unknown was called—was understood, and therefore left out of the notation altogether. In 1591, François Viète would have written that same polynomial as *A quad − 3 in A+9 plano*. Then in 1631, Thomas Harriot would have written it as $xx - 3x + 9$. He also did something extremely clever. Until that point in time, mathematicians were interested in the roots of polynomials—that is, in the numbers that would make the polynomial equal to zero. Harriot had the ingenious idea of first setting the polynomial equal to zero, thereby setting up an equation, a polynomial equation, and asking for the numbers that satisfy the equation. You might think this is only a grammatical difference, and that it could not lead to anything new. But it opened a door to a whole new way of thinking about polynomials. He saw that a polynomial could be built up from a product of its factors, just as a number could. For example, as we will see later that
$$x^4 - 4x^3 - 19xx + 106x - 120 = (x - 2)(x - 3)(x - 4)(x + 5).$$

This ingenious idea was a game-changer—the problem of finding the roots of polynomials quickly became the problem of factoring polynomials. It was a slight grammatical maneuvering that advanced the possibility that every polynomial equation has a root, possibly a real number or a complex number. Proving it generally with satisfactory rigor was beyond seventeenth-century means. Such a general proof would have to wait almost two hundred years, when it would take on the general form that we now know as the *fundamental theorem of algebra*, which tells us that any polynomial has as many roots (counting multiple ones) as its highest power.

Even more significant: the idea gave mathematicians a sense of proper algebraic form. By putting all terms on the left of an equation and leaving an isolated zero on the right, the form comes to the forefront. The equation $x^2 + 3x + 2 = 0$ falls into a category recognizably distinct from, say, $3x + 2 = 0$, and the equation $5x^3 + 2x^2 + 3x + 2 = 0$ falls into a third category. Algebra was not just about equations, but also about *forms*.

And finally, in 1637 René Descartes had the idea of using numerical superscripts to mark positive integral exponents of a polynomial in his *La Géométrie*, a work that could be read easily by anyone tuned to our modern notation. That simple idea of ranking the individual powers numerically, which seems obvious to us modern folk who see the powers as a counting of the number of times the variable is multiplied by itself (that is, x^2 as $x \cdot x$ and x^3 as $x \cdot x \cdot x$), at once transformed the way we see and work with polynomials. Descartes was extending the tradition of writing x for the unknown and xx for x^2 that was started by Viète and Harriot, to a ranking scheme.

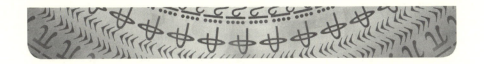

Chapter 15

Vowels and Consonants

François Viète, who wrote under the Latin name Franciscus Vieta, was a French mathematician having the great advantage of working in an era of abundant mathematical contributions from the Italians, Germans, and English.

He expressed his famous computation for π entirely in six paragraphs of 139 words in proposition II of his *Isagoge*.[1]

> Propositio II
>
> Si eidem circulo inscribantur polygona ordinata in infinitum, & numerus laterum primi sit ad numerum laterum secundi subduplus, ad numerum vero laterum tertii subquadruplus, quarti suboctuplus, quinti subsexdecuplus, & ea de inceps continua ratione subdupla.
>
> Erit polygonum primum ad tertium, sicut planum sub apotomis laterum polygoni primi & secundi ad quadratum à diametro.
>
> Ad quartum vero, sicut solidum sub apotomis laterum primi secundi & tertii polygoni ad cubum à diametro.
>
> Ad quintum, sicut plano-planum sub apotomis laterum primi secundi tertii & quarti ad quadrato-quadratum à diametro.
>
> Ad sextum, sicut plano-folidum sub apotomis laterum primi secundi tertii quarti & quinti polygoni ad quadrato-cubum à diametro.
>
> Ad septimum, sicut solido-solidum sub apotomis laterum primi secundi tertii quarti quinti & sexti polygoni ad cubo-cubum à diametro. Et co in infinitum continuo progressu.

His proposition II tells us how to approximate π by first inscribing a square in a circle, projecting the bisection of each side out to the circle to get an octagon, and repeating the process, first with the octagon, and then again and again with each resulting polygon.[2] It is an old method of Archimedes, tweaked a bit to make the calculations simple. Viète ends his proposition with the six-word sentence, *Et eo in infinitum continuo progressu,* which translates to "And we continue it progressively into the infinite." This is the first time (as far as I know) where any European author has used this idea of continuing an algebraic process indefinitely. In the end, we find π by finding $\frac{2}{\pi}$ to be equal to the following infinite product of infinitely nested terms:[3]

$$\frac{2}{\pi} = \frac{\sqrt{2}}{2} \cdot \frac{\sqrt{2+\sqrt{2}}}{2} \cdot \frac{\sqrt{2+\sqrt{2+\sqrt{2}}}}{2} \cdots$$

Even Rudolff and Chuquet had no proper notation for expressing such an infinite sum of nested square roots, though conceivably some of the earlier German manuscripts could have.

Early in his career Viète would have written the polynomial equation $x^2 - 3x = 2$ as *quadratum in A, minus A ter aequetur 2,* where the *A* represented the unknown that we would designate as x. At other times, he would have used + and − to symbolize plus and minus in order to write that same equation as *quadratum in A,−A ter aequetur 2.*[4]

Later, he wrote, *X quadratum in A ter, minus A cubo, aequetur X quadrato in B.*[5] This translates to $3X^2A - A^3 = X^2B$ (or, what we would write as $3a^2x - x^3 = a^2b$), which is the equation one gets from trying to trisect a given angle embedded in a circle of radius X whose chord is B. The A represents a chord that is the unknown third of the angle. Here, the variable is A, not X.

Viète was showing us an intimate link between Greek geometry and algebra, a link from the mathematics of lines, figures, and solids to the underlying channels of symbolic algebra, a link that had been there all along but hardly ever fully appreciated. Yes, there had been commentators who saw the links very clearly. Heron of Alexandria figured an algebraic approach to Euclid's geometry in the first century, and Petrus Ramus wrote about the connections between geometry and algebra in his *Twenty Seven Books of Geometry* in 1569.[6] But it was Viète who made them clearer than they had ever been.[7]

In his *Geometria*, Viète was interested in equations of the form $3X^2A - A^3 = X^2B$. They are what are now called "homogeneous equations," where the sum of the powers of the letters in each term is 3. The idea was to add terms only of the same dimension. Equations of such form appear in Euclid's *Elements*. Book II is all about the geometry of rectangles, squares, and other figures. If you are wondering how geometry could be algebra, think of it this way: the operation of addition or subtraction is the same as extending or cutting off lines; the product of two numbers a and b is the same as the geometric construction of a rectangle having adjacent sides a and b. Extracting the square root of a is the same as finding a square whose area is a. We are able to see this from an algebraic point of view; however, Euclid was proving a geometric theorem, not an algebraic one.

Inspect proposition 4 from *Elements*, book II, for hints of algebra.[8] It says:

> If a straight line be cut at random, the square on the whole is equal to the squares on the segments and twice the rectangle contained by the segments.

Though this is a translation into English, the meaning may seem foreign. The phrases *square on the whole*, *squares on the segments*, and *rectangle contained by the segments* need interpretation. For now, trust that their meanings will become clear on further examination.

Mark the two ends of the line as A and B, and the random cut as C. We use the convention that any line whose ends are marked by letters—say, A and B—will be labeled AB. Also, any rectangle whose four corners are each marked by a letter will be labeled as a juxtaposition of those four letters. Construct a square with side of length AB. Label the corners of that square A, B, E, D, as they are in figure 15.1, and draw a line CF perpendicular to AB at C.

Now we can interpret *square on the whole* to mean the square whose side is the whole (original) line AB; that would be the square $ABED$ (see figure 15.1). Likewise, the *square on a segment* means the square whose side is a segment of AB. There are two such squares; one is $HGFD$ and the other is $CBKG$. We take the *rectangle contained by the segments* to mean the rectangle whose sides are the lengths of the

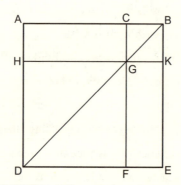

FIGURE 15.1 The square on the whole.

two segments cut by C; that rectangle is $ACGH$. From the illustration, we see that the area of $ABED$ is equal to the area of $HGFD$ plus the area of $CBKG$ plus the area of $ACGH$ plus the area of $GKEF$.

Just by letting $a = AC$ and $b = CB$ and looking at that last sentence algebraically, we find that

$$(a + b)^2 = a^2 + b^2 + 2ab.$$

It is an identity that we can prove from just a few basic laws of arithmetic.[9]

But Viète was still using words and sentences rather than abbreviations and symbols to work in his version of algebra. It seems that he preferred the words, even though he was familiar with Recorde's, Bombelli's, and Stevin's notation. His great contribution to algebra was not an introduction of new operational symbols. There are virtually no new operational symbols in Viète's works; rather, it was the abstract use of letters to represent the more general nature of the objects in play, together with the magnificent idea that those letters were also to be subject to algebraic reasoning and rules just as much as numbers. Even in al-Khwārizmī's time, it was known that common factors in an equation may be cancelled. Viète generalized the notion of cancellation to let us know that if, say, $BE^2 + B^2E = B^3$, then $E^2 + BE = B^2$, and also that if, say, $BE^3 + 3B^2E^2 = 3E^4$, then $BE = 3B^2 = 3E^2$. In other words, nonzero unknowns are canceled by the same rules as those governing the cancellation of knowns.

Why was that not done before? One answer is that it would have been confusing to write the polynomial almost entirely in letters. We write the quadratic polynomial as $ax^2 + bx + c$ and can immediately distinguish the unknown quantity from the specifically known. Earlier notation did not provide such a clear distinction between knowns and unknowns. A second answer is that the performance of arithmetic operations (addition, subtraction, multiplication, division, and root extraction) on abstract magnitudes seemed to be merely symbolic gestures that had no computational advantage. To write $3 \times 4 = 12$ is a comfortable operation that groups three things with four things. But what could be the advantage of writing *B multiplicata per C aequalis B multiplicata per C*? It's like giving the suspiciously redundant tautology $B \times C = BC$.

Viète used vowels to represent unknowns, and consonants to represent knowns. That convention does two things: it avoids confusion between representations of knowns and unknowns and—more importantly—permits us to have multiple unknowns. An equation could then distinguish between the unknowns. For instance, the equation $3X^2E - U^3 = X^2B$ is about two unknowns, E and U, though we would write the equation as $3a^2x - y^3 = a^2b$, where a and b are known constants, and x and y are unknowns.

There is another advantage of Viète's lettering system. The vowel A was used to mark the unknown quantity, and the successive powers were then marked as *A quad.*, *A cubus*, *A quad. quad.* Like Chuquet's system, where the roots were marked—and therefore ranked—as $\mathbb{R}^1, \mathbb{R}^2, \mathbb{R}^3, \mathbb{R}^4 \ldots$—Viète's system provided a mental connection between the ranked powers of a single unknown, in just the same way our notation $x, x^2, x^3, x^4, \ldots$ does. Though no new operational symbols were used in all of Viète's writing, the introduction of his system of letters was almost essential to further advancement in mathematics.

Today, we may think Viète's lettering system is such an obvious notational convenience, but even as late as the end of the sixteenth century, such a notion was revolutionary. For that, he too is sometimes accepted into the fathers of algebra family.

Viète's vowel-consonant notation had a short life, but it inspired a huge advance for symbolic algebra. It seems hard for us to conceive such an idea as being so clever. For us, it is natural for letters to represent fixed known numbers and variable undetermined numbers. But we are products of intelligent habit. We learn things and forget that there was ever a time when we had to learn them to know them. We, who live in the second decade of this twenty-first century, may now find cell-phones and GPS navigation technologically sophisticated, but in the next century, when such phenomena will be replaced by even more advanced phenomena, people will look back and think of our advances as we now view the simplicity of, say, the typewriter, or the garden hose. Immersed in our own common use of symbols, it is difficult to conceive of why Viète's idea had not occurred to so many of the brightest mathematicians from Diophantus to Bombelli.

One might think that there is hardly a conceptual difference between Rudolff's $\mathbb{R}$, Fibonacci's *res*, Chuquet's $\mathbb{R}$, and Viète's vowel-consonant notation. But there is. Viète's vowels neither drag along the taboos of culture nor restrain perceived notions of what number is supposed to be. Both $\mathbb{R}$ and *res* mean what they say. Even Chuquet's $\mathbb{R}$ means what it says. They are not symbols in any true sense, since they carry preconceptions representing things that they resemble. Viète has given us something more than merely new notation. His *A* (our *x*) is a true symbol; it transcends the concreteness of the object it is assumed to represent. Tobias Dantzig once wrote, "It is this power of transformation that *lifts algebra above the level of a convenient shorthand*."[10]

But there is another benefit to the vowel-consonant notation. The advantage that comes from being able to perform operations on Viète's vowels and consonants while transforming awkward literal expressions into more convenient equivalent forms. It is that advantage of transformation, Dantzig again tells us, that "lifts algebra above the level of a convenient shorthand."

And still there is another feature: the vowel-consonant notation that lifts algebra. Imagine what algebra would be like if instead of dealing with the general expression $ax^2 + bx + c$, we had to specify what the coefficients a, b, and c were as numbers. It would mean that any solution to a problem involving, say, the individual quadratic polynomial $x^2 + 2x + 3$ would be thought of as different from, say, the individual quadratic polynomial $2x^2 + 3x + 1$, even though the first polynomial would quickly suggest a concrete procedure for solving the second. Each expression would have to be handled differently, though of course there would be clues on how to proceed. Viète's wonderful vowel-consonant notation gives us a way of contemplating and working with the collective, the general, *the any*, and *the all*. This too, Dantzig would agree, is the power that "lifts algebra above the level of a convenient shorthand."

More significant is the role it played in forming the generalized number concept. Before Viète, algebraists would have seen $x^2 + 2x = 3$, $x^2 - 2x = 2$, and $x^2 - 2x + 2 = 0$ (using our x-for-the-unknown notation) as distinct types of quadratic equations, which, of course in a certain sense, they are. It was assumed that an expression of the form "square and something times an unknown equals a number" is not the same thing as "square equal to something times an unknown and a number." That was not at all due to any timidity in manipulating symbols, but rather due to the troubles and dilemmas concerning negative numbers and zero.[11] We see all three at once to be of the same form: $x^2 + bx + c = 0$.

The first equation is satisfied by $x = 1$, the second by … well, it doesn't seem to have a rational solution, and neither does the third. We know now that the second equation has two solutions: $x = 1 + \sqrt{3}$ and $x = 1 - \sqrt{3}$. When we try to find the solutions to $x^2 - 2x + 2 = 0$, we end up with symbols that look like $x = 1 + \sqrt{-1}$, and $x = 1 - \sqrt{-1}$.

These last two solutions had no meaning in Viète's lifetime. But when expressed in general terms with notation such as $x^2 + bx + c = 0$, we find that the solutions are always

$$x = \frac{-b \pm \sqrt{b^2 - 4c}}{2}.$$

That general solution called for a distinction by species.

1. Those solution candidates where $b^2 - 4c$ is a perfect square: perfectly acceptable solutions.

2. Those solution candidates where $b^2 - 4c$ is not a perfect square and $b^2 > 4c$: suspicious solutions. Still not accepted as valid, although on the verge of acceptance.[12]

3. Those solution candidates where $b^2 < 4c$: completely meaningless solutions, *the complex numbers*, whose existence as numbers were denied.

Here is De Morgan on the topic:

> [Viète] concluded that subtraction was a defect, and that expressions containing it should be in every possible manner avoided. "Vitium negationis," was his phrase. Nothing could make a more easy pillow for the mind, than the rejection of all which could give any trouble;... The next and second step,... consisted in treating the results of algebra as necessarily true, and as representing some relation or other, however inconsistent they might be with the suppositions from which they were deduced. So soon as it was shown that a particular result had no existence as a quantity, it was permitted, by definition, to have an existence of another kind, into which no particular inquiry was made, because the rules under which it was found that the new symbols would give true results, did not differ from those previously applied to the old ones.... When the interpretation of the abstract negative quantity showed that a part at least of the difficulty admitted a rational solution, the remaining part, namely that of the square root of a negative quantity, was received, and its results admitted, with increased confidence.[13]

De Morgan may not have had it quite right from a modern historian's point of view. These strange species were known long before Viète's time. The Pythagoreans encountered irrational numbers soon after thinking about squares and right triangles, and Cardano had timidly pondered complex numbers in his *Ars Magna* in 1545. But Viète's notation brought those "true" and "fictitious" roots closer to the surface because the general notation exposed one important fact: that they had relevance as intermediate solutions to real problems—that is, somehow, the algebraic solutions gave correct answers, even though they involved meaningless steps.

They may have been meaningless solutions in the sixteenth century, but in the seventeenth century, with more general notation, more attention was paid to the meaningless than had ever been paid before. So that attention called out the question: What is number? A fundamental question. But it was also a deep one, far too deep for calling before its need. Our more sophisticated concept of number now accepts the square root of a negative number into the family. That enrichment gave us the *fundamental theorem of algebra*, which tells us that any polynomial of any degree $n \geq 1$ always has n roots that may or may not be distinct.[14] Of course, those roots may be (and are likely to be) complex numbers. Why fundamental? Two reasons at least: (1) because it tells us that every polynomial is just a product of degree-1 polynomials, each of the form $(x - r)$, where r is a root; and (2) because it guarantees an answer to every question that leads to a polynomial.

And Viète's more general notation called attention to another question: What is form? The equation $ax+by+c = 0$ is almost all letters. The letters a, b, and c represent known values, whereas the letters x and y take on a whole range of unknown values. We think of a, b, and c as representations of values, without any interest in what they actually are. Thus, the entire equation is first and foremost thought of as a relationship between x and y. But once that relationship is established, this marvelous understanding of notation permits us to further examine the form $ax + by + c = 0$ by varying the values of a, b, and c (so-called parameters) and thereby set up a family of relationships between x and y. The form of an equation then becomes a new object of study, one that leads to a classification of equations that could not have been dreamed of without the symbolic distinction between the two sets of values, constants and variables.

Viète finished his *Isagoge* with the last four words all in capitals, and put his pen down for the final full-stop punctuation mark. He wrote:

> *Denique fastuosum problema problematum ars Analytice, ... jure fibi adrogat, Quod est, NULLUM NON PROBLEMA SOLVERE.*[15]

> Translation: Finally, the analytical art,...appropriates to itself by right the proud problem of problems, which is, TO LEAVE NO PROBLEM UNSOLVED.

Chapter 16

The Explosion

René Descartes's *Geometria* was published just thirty-four years after Viète died. It had a new idea for notation, a rule: beginning letters of the alphabet were to be reserved for fixed known quantities and latter letters (past p) were to represent variables or unknowns that could take on a succession of values. Descartes seems to have followed Thomas Harriot's practice of using lowercase letters, though he denied ever having seen Harriot's writings. To this day, this division of the alphabet at p remains the loose standard rule.

The German philosopher Daniel Lipstropius, a contemporary and biographer of Descartes, told us that Descartes's most brilliant idea came to him while watching a fly crawl along a curved path. It was a fable of course, implying that the Cartesian coordinate system owes its origin to Descartes describing the path in terms of its distance from the walls, that a fly was responsible for one of math's most radical shifts: a relatively early marriage of algebra and geometry. It was a fable because Descartes's coordinate system looked nothing like our modern one with its horizontal and vertical axes indicating related variables.[1] The story later inflated into a more sweeping fiction of how Descartes, because of his poor health, would lie in bed late each morning meditating on how all of science could be made as certain as mathematics.

If the fly traced a curved path in space, it would also have left a trail of arithmetical data, and Descartes would have understood that the geometry of the curve could be reconstructed from the arithmetical data and, conversely, that the arithmetical data could be reconstructed from the geometry of the curve. Geometry and arithmetic were simply different interpretations of the same mathematics: algebra and geometry are intimate echoes of each other. Miraculous!

It's true that Descartes was in the habit of lying in bed till late morning thinking about his surroundings and existence. As a boy, he was permitted to stay in bed to nurse his uncontrolled coughs, which seemed to fade by afternoon. Poor medical advice for someone probably suffering from postnasal drainage that would have been relieved more quickly by getting out of bed. Nevertheless, he probably thought about the physical world, how it is fundamentally mechanical, how everything in nature can be explained through the laws of mechanics, how all of theoretical physics should be expressible through a small number of general laws and observable facts of nature, and how a small number of principles and fundamental equations, could be expressed through algebraic equations.

There is that wonderful Pythagorean theorem that tells us that there is a structural relationship between three squares sharing sides with a right triangle. How is it that that theorem gives us an easy way of finding the distance between any two points in space? And how is it that we are able to represent straight lines and conic sections (ellipses, parabolas, and hyperbolas, those curves marked by a plane cutting through a cone) by equations and proportions?

The intimate link between geometry and algebra had been suspected since Plato's time, when mathematicians of the Academy worked on trisecting angles, duplicating cubes, and squaring circles. In the third century BC, Apollonius of Perga investigated curves that could be produced by cutting a cone by a plane—the ellipse, the parabola, and the hyperbola. Heron of Alexandria figured out an algebraic approach to calculating surfaces and volumes in the first century. The geometer Pappus, a contemporary of Diophantus, hinted that there should be some connection between geometry and algebra. Menaechmus, in the fourth century, discovered connections between conic sections and equations, and early Greek geographers made free use of coordinate systems. Nicole Oresme, in 1361, worked with a system of latitudes and longitudes introducing early ideas of a coordinate system, complete with a horizontal line to represent time and a vertical line to represent speed.

Geometry had its origins in the interest of working with lines, figures, and solids that could be imagined in the mind. Algebra had its origins in problems involving number—number hinged by geometric conceptions of iconic figures. By the late Middle Ages, algebra was progressively focusing on more abstract notions of number, especially after Viète advanced the notation to include constants and unknowns, a notation that liberated algebra from the confines of geometric metaphor. It could leap toward its more general purposes, a concentration on abstract magnitudes.

We saw that the algebraic operations of addition, subtraction, multiplication, division, and extraction of a square root all have matching operations in geometry. But could those operations actually be performed? Viète knew that the product of two numbers a and b is the same as the geometric construction of a rectangle having adjacent sides a and b, and that extracting the square root of a is the same as finding a geometric square whose area is a. But how could it be done, actually?

Descartes showed us how to on the second page of his *Geometry*.[2] First, multiplication: suppose we have two line segments labeled AB and AC. Place them in any way or place, but have them joined at end A. (See figure 16.1.)

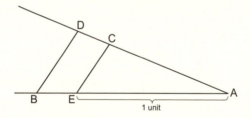

FIGURE 16.1 Multiplication.

On AB, mark off a line segment of one unit, and label it AE. You may have to extend AB if it is shorter than one unit. Connect E to C, and construct line BD parallel to EC. From the similar triangles we see that AD is to AB as AC is to 1, and therefore $AD = AB \times AC$.

To divide, use the same set up; notice that the quotient AD/AB must equal AC.

To find the square root of AB, extend AB by one unit to C. (See figure 16.2.)

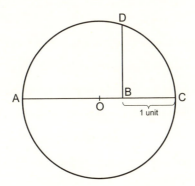

FIGURE 16.2 Root extraction.

Bisect the line AC at O. Construct a circle of diameter AC centered at O. Draw a line at B perpendicular to AC. It meets the circle at D. Then $BD = \sqrt{AB}$.

All these operations are constructible with straightedge and compass, and therefore provable from Euclid's axioms. And any problem that can be expressed through a geometric construction that uses straightedge and compass alone can also be expressed by a polynomial equation of degree one or two.

The Cartesian coordinate system is more than just an orienting system, more than just a way to get from here to there. It is a way to see geometry through the lens of algebra. Descartes (and Fermat too) gave us something incredibly special. He showed us that thinking itself has optional modes. He taught us that we have optional modes for conceptualizing problems. We may wish to attack the question of whether or not an arbitrary angle can be trisected using straightedge and compass alone, an ancient problem of geometry. It may be naturally expressed geometrically with words such as "line" and "angle." But sometimes we are fooled by what we think is natural, entrapped and limited by unnecessary contortions of conceptualization.

Descartes gave us a way to switch between modes of conceptualizing, to translate geometric problems to an algebraic coordinate system. Points, lines, and curves of the Greek geometers were free to be represented by abstract algebraic expressions, freed from the shackles of our physical impressions of space, enabling imagination to wander far beyond the tangible world we live in, and into the marvels of generality.

Trisecting an arbitrary angle turns out to be a question about whether or not a rational root to a particular cubic equation exists. Descartes could not have known the answer, which we now know: the root in question does not exist.

To see how the Cartesian system gives us the link between geometry and algebra, we briefly remind ourselves what we either learned or missed in high school math. Keep in mind that the coordinate system invented by Descartes (as well as by Fermat) was not quite the system we use today. In fact, it was not the first idea of a coordinate system: the fourteenth-century cleric Nicole Oresme had a similar idea. But Descartes's idea was a kick-starter for today's more developed concept that was introduced later. We start by viewing the world on a flat plane with a fixed benchmark chosen arbitrarily, just as you might fix a tall building to orient a walk through an unfamiliar city. We label the benchmark $(0, 0)$, a symbol whose design will soon become clear. (See figure 16.3.) The flat plane, like the flat surface of a computer screen, has horizontal and vertical number lines through $(0, 0)$, where distances in the directions of the arrows are positive, and distances in the opposite directions are negative.

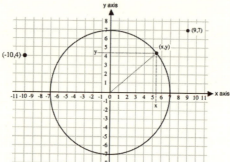

FIGURE 16.3 The address of a point.

Pick any point in that flat plane. For the moment, call it P, and ask where P is in relation to our benchmark $(0, 0)$. One natural description (though there are others) for P's position is how far it is (positively or negatively) in relation to the arrows (using whatever units you wish) in the horizontal and vertical directions from $(0, 0)$. For example, $(9, 7)$ is the address of the black point in the upper-right corner, and $(-10, 4)$ is the address for the black point at the left of the circle.

To represent a circle of radius 7 centered at $(0, 0)$ using this wonderful system, all we have to do is describe an arbitrary point (x, y) on the circle as one end of a line joined to $(0, 0)$, which is after all the radius of the circle and therefore always equal to 7 units in magnitude. By the Pythagorean theorem applied to the triangle of base x and height y, we have $x^2 + y^2 = 7^2$. Every pair of numbers x and y that satisfies the equation $x^2 + y^2 = 7^2$ will give a coordinate address of a point on that circle of radius 7. Conversely, any point on that circle will have a coordinate address (x, y), where x and y satisfy the equation $x^2 + y^2 = 7^2$. We used the circle merely as an example to show how surprisingly simple the linkage is.

It would seem that Descartes would have confidently used the equations of his analytic geometry to solve problems of geometry. He solved them, but always found a need to confirm his algebra proofs by geometry.[3] Newton and Leibniz would do the same in their infinitesimal calculus. It may have been simply that they were holistic mathematicians who wished to see all the strings.

Surprising as it may seem (beyond the convention of using lowercase letters), using beginning letters of the alphabet for fixed known quantities and latter letters for unknowns, Descartes invented very few new symbols. He introduced a tweaking of Bombelli's and Stevin's indexing of powers of the unknown, and used superscripts to show the numerical powers of the unknown. Oh—and, yes—there is the matter of the vinculum, that horizontal bar joined with the old German symbol for square root $\sqrt{}$ to indicate that all terms beneath the bar are to be grouped together before the root is extracted. It is our modern symbol for square root $\sqrt{\quad}$. We have seen how important an advance it is.

In *Geometria*,[4] Descartes writes:

> Et aa, seu a^2, ad multiplicandam a in se; Et a^3, ad eandum adhuc semel multiplicandam per a, atque ita in infinitum; Et $sqrta^2 + b^2$, ad extrahendam radicem Quadratam ex $a^2 + b^2$; Et $\sqrt{C.a^3 - b^3 + abb}$, ad extrahendam radicem Cubicam ex $a^3 - b^3 + abb$, & sic de cæteris.

Translation: And aa, or a^2 to multiply a by itself; And a^3, once more multiplied by a, and so on indefinitely; And $\sqrt{a^2 + b^2}$, for the extraction of the square root of $a^2 + b^2$; And $\sqrt{C.a^3 - b^3 + abb}$ for the extraction of the cube root of $a^3 - b^3 + abb$, and so on.

Here, we have the blending of the German radical sign $\sqrt{\ }$ with a vinculum to cover the expression whose root is to be extracted. Our current symbol $\sqrt[3]{\ }$ for cube root would not appear for another thirty years, when it appeared in several places at once: in Michel Rolle's *Traité d'Algèbre* and in a letter from Gottfried Wilhelm Leibniz to Pierre Varignon.[5]

Then, on page 4 of Descartes's *Geometria*, we find polynomials written the way we would, except for that strange symbol ∞ that Descartes would use to write "is equal to."[6] The z-like figure is just a flamboyant script z.

> $Z \infty 6$, *aut*
>
> $Z^2 \infty - aZ + b^2$, *aut*
>
> $Z^3 \infty + aZ^2 + b^2Z - c^3$, *aut*
>
> $Z^4 \infty + aZ^3 + b^2Z^2 - c^3Z + d^4$, *&c.*

On page 69, for the first time, we find a perfectly readable account of the equation that almost looks as if it is out of a twentieth-century textbook.[7]

> Sciendum itaque, quòd icognita quatitas in qualibet Æquatione, tot diversas radices seu diversos vatlores habere profit, quot ipsa habet dimensiones. Nam si, exempli gratiâ, x supponatur æqualis 2, seu $x - 2$ æqualis nihilo; & rursus $x\infty 3$, seu $x - 3\infty 0$; & multiplicetur $x - 2\infty 0$ per $x - 3\infty 0$; habebitur $xx - 5x + 6\infty 0$, seu $xx \infty 5x - 6$. quæ Æquatio est, in qua quantitas x valet 2, & præterea etiam 3. Quòd si rursus fiat $x\infty 4$, atque $x - 4\infty 0$ multiplicetur per $xx - 5x + 6\infty 0$, producetur $x^3 - 9xx + 26x - 24\infty 0$. quæ alia eft Æquatio, in qua x habens tres dimensiones, tres quoque habet valores, qui sunt 2, 3, & 4, atque una falsa, quæ es 5.

Descartes writes that if we multiply the polynomial $x - 2$ by $x - 3$, the result is $x^2 - 5x + 6$. If we multiply that by $x - 4$, we get $x^3 - 9xx + 26x - 24$, and if we continue and multiply that last polynomial by $x + 5$, we get $x^4 = 4x^3 - 19xx + 106x - 120$. So the roots of the polynomial $x^4 = 4x^3 - 19xx + 106x - 120$ are 2, 3, 4, and -5, and therefore $x^4 = 4x^3 - 19xx + 106x - 120$ and $(x-2)(x-3)(x-4)(x+5)$ are just two different representations of the same polynomial.

In the preceding chapter, we saw how Viète's introduction of consonant letters of the alphabet advanced algebra. Before letters were used for that kind of representation, the polynomial notations of Chuquet, Bombelli, and Stevin were perfectly adequate. The square or cube of a known number could be calculated to get another known number; there is no need to write 2^2 when you just mean 4. In any polynomial, only the unknown quantity was raised to a power, and that unknown-to-a-power could not be directly calculated. So, for sixteenth-century manuscripts, it was fine to write 3^2 for $3x^2$, to write 3^1 for $3x$, and to write 3^0 for 3. Chuquet, Bombelli, and Stevin used such a so-called *Index Plan* for writing exponents without ambiguity.

However, there was a problem. A polynomial could have more than one unknown, an x and a y, for instance; so $3x^2 + 5y^2$ could not be written under the *Index Plan*. We favor Descartes's notation mostly because it is our own, but also because it is preferable to the ambiguous notations that came before it, and also because nobody has come up with anything better, ... yet.

A symbol may become conventional and remain in use for centuries, until something happens to advance a context in which it appears to create obstructions. It happened to the sixteenth-century Index Plan. Polynomial algebra notation now has a seasoned notation that is unlikely to change over the next millennium. Pacioli's ℞ lasted on and off in some places for close to two hundred years. Rudolff's √ had some minor tweaks as it competed with other less worthy attempts to mark square roots, but didn't change for a hundred years, until Descartes added a vinculum.

In Descartes's *Geometria*, we see our very own square root symbol (figure 16.4 top) as the German symbol for square root √ with a vinculum added to unify the terms whose root should be extracted.[8] Some people must have thought that the ℞ was there to stay, just as we now think that our exponent notation is here to stay.

$$Z \propto \frac{1}{2}a + \sqrt{\frac{1}{4}aa + bb}.$$

Quòd si verò habeatur $yy - ay + bb$, atque y sit quantitas, quam invenire oportet, facio rursus idem triangulum NLM, & à base ejus MN ausero NP, æqualem NL, eritque reliqua PM, æqualis y, radici quæsitæ. Ita ut fiat $Z \propto -\frac{1}{2}a + \sqrt{\frac{1}{4}aa + bb}$. Nec aliter sit, si proponatur $x^4 \propto - ax^2 + b^2$. PM enim esset x^2, & haberetur $x \propto \sqrt{-\frac{1}{2}a + \sqrt{\frac{1}{4}aa + bb}}$: atque ita de aliis.

FIGURE 16.4 The vinculum in René Descartes, *Geometria* (1659), page 3.

There is a nesting of square roots, from which, without much imagination, we can foresee a continued nesting that could be used to show off Viète's wonderful proposition II (in chapter 15) that approximates $\frac{2}{\pi}$.

Florian Cajori tells us that Descartes introduced the new radical sign, vinculum and all. The curious puzzle is this: who actually came up with the idea of the vinculum? When Francisci van Schooten edited Viète's *Opera mathematica* in 1648, he already used the vinculum in his commentaries. Hmm,... the figure (figure 16.4) is from a page of van Schooten's edition of Descartes's *Geometria*, published in 1659. Could van Schooten have slipped the vinculum into the *Geometria* to simplify Descartes's meaning?

Fortunately for us, Descartes had great influence as a mathematician and was able to help standardize the best notation going into the next century. The seventeenth century was filled with experiments using all sorts of odd and cumbersome notation that could have impeded progress in mathematics for years.

For Greek geometers, a curve was more or less a static figure. Descartes was beginning to think of a curve differently. His coordinate system was thought to be a collection of dynamically moving points determined by a rule (its equation), an algebraic object with addresses (that is, points) indicated by real numbers x and y. Those real numbers, the "coordinates," were locked together in a co-ordered numeric relationship; one could not change without the permission of the other. This new geometry looked at curves as relations between variables. It was a very great advance, one that radically changed the tactics and manner of mathematics, one that made calculus possible, and one that changed forever how we think of motion.

A typical early-seventeenth-century empirical observation would have shown the height of a projectile at various times as a table of values. There were no clues for finding heights at times when the projectile was not observed. With the notion of a graph, and an algebraic equation relating a height h to any time t, came an intuitive understanding of how height changes smoothly as time changes, a picture of how the numbers were climbing or descending.

The unity of geometry and algebra was one of the greatest discoveries. At once, it gave a picture of the law governing an event as well as the connection between dependent events. It gave later mathematicians the powers needed to picture and articulate mathematically how, in two related phenomena, a change in one affects a change in the other.

Still, scientists were divided over the question of whether nature was fully mechanical and fully explainable by mathematics.[9] But this new kind of union of geometry and algebra suggested that the secrets of the universe could be fully explained mathematically. Space and time were linked, not only through indefinite, unreliable geometric pictures caught by the spirit of intuition, but also through algebra.

The concept of a function would have been natural for examining the space-time relationship, but that would have to first wait till 1692 for Leibniz to introduce a proto-concept and then, after some tweaking by Johann Bernoulli and Leonard Euler, wait again until 1834 for Gustave-Peter Lejeune Dirichlet to introduce his version.[10]

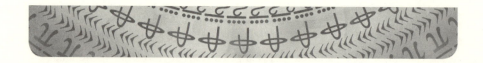

Chapter 17

A Catalogue of Symbols

William Oughtred died on Sunday, the thirteenth of June 1660, at the age of eighty-eight. John Aubrey tells us, "He was a little man, had black haire, and blacke eies (with a great deal of spirit). His head was always working. He would drawe lines and diagrams on the dust. He had burned all his papers, claiming that, 'the world was not worthy of them.' He was so superb. He burned also several printed books, and would not stirre, till they were consumed."[1] If you examine the engraved portrait of him by the Czech engraver Wenzel Hollar, you will find a thin-lipped man of late life with an immense nose that begins at the top of his eyebrows (figure 17.1).

GULIELMUS OVGHTRED ANGLVS.
ex Academia. Cantabrigiensi Æ ætat 78. 1646.

FIGURE 17.1 William Oughtred.

He finished the first edition of his *Clavis mathematicae* in 1631. It went through many editions and was a popular teaching book for a half century after his death. In the *Clavis*, we find the first use of the St. Andrews cross × as a symbol of multiplication. It had been used in medieval times as an indicator of many things other than the product of two numbers. Until Oughtred's *Clavis*, multiplication was signified by juxtaposition; *ab* meant *a* times *b*. That was fine, as long as the multiplicands were symbols themselves. When it came to definite numbers, there was ambiguity. Did 22 mean the number *twenty-two*, or 2 times 2?

The use of juxtaposition was not a symbol; it was a concept of notation that tempted confusion. In 1545, Michael Stifel used the letters M and D for multiplication and division, respectively. So did Simon Stevin in 1585. They would write 3②D *sec* ①M *ter* ② to indicate

$$\frac{3x^2z^2}{y},$$

where *sec* stood in for "second unknown" and *ter* for "third unknown." Once again, the M, D, *sec*, and *ter* are not true symbols, merely abbreviations.[2] They are susceptible to conceptual confusion: which unknown is the first, which the second, or which the third? Our present notation, thanks to Descartes, avoids that problem because the letters of the alphabet are already ranked by order.

Viète wrote "A in B" to mean the product of A and B. As late as the turn of the twentieth century, some authors were using M to indicate multiplication.[3] And even today, there are ambiguities attributed to juxtaposition: we write $3\frac{1}{2}$ to mean $3 + \frac{1}{2}$. Perhaps that is why so many young students make so many errors when computing with mixed fractions.

Of the more than one hundred prospective symbols and labels introduced by Oughtred, less than a dozen are still in use. Still, anyone who can design half a dozen symbols well enough to become standards that survived for more than three centuries deserves applause.

By the seventeenth century, most rhetorical mathematics writing shifted to symbolic writing. All sorts of new notation were introduced, some useful, some not, some impractical, and some downright foolish. But progress continued. In the preface to his *Cursus mathematicus*, published in 1634, Pierre Hérigone wrote, "I have invented a new method of making demonstrations, brief and intelligible, without the use of any language." He meant that he had introduced a complete system of mathematical notation. Yet the only symbols of his complete system that are still in use today are his geometrical ones, ⊥ ("is perpendicular to") and ∠ ("angle").

Alfred North Whitehead once wrote, "There is an old epigram which assigns the empire of the sea to the English, of the land to the French, and of the clouds to the Germans. Surely it was from the clouds that the Germans fetched + and −; the ideas which these symbols have generated are much too important for the welfare of humanity to have come from the sea or from the land."[4]

The letters *p* and *m* replaced the words *plus* and *minus*. Popular history attributes the signs + and − to Stifel. But there is also evidence that Stifel saw those signs elsewhere. It has been suggested that they may have first appeared as chalk marks on inventory in German warehouses to designate excess or defect from a standard weight.[5]

They appeared in Stifel's 1544 edition of his *Arithmetica Integra*. They also appeared in Johannes Widmann's 1489 work, *Behende und hubsche Rechenung auff allen Kauffmanschafft* (*Nimble and Neat Calculation in All Trades*). Widmann's +, however, was not the addition operation; rather, it meant "excess," as in "+2 is two more than what was expected." For a time afterward, there were competing marks for the addition operation. A favorite was the abbreviation *p*, or *p̄*, the line through or above *p*, to distinguish the operation from a quantity. Tartaglia preferred to use the Greek letter φ (phi) to indicate addition. Symbols for minus date back to Diophantus's time, when it looked like an arrow pointing upward or downward. The Latin cross, oriented horizontally as ⊹ was popular, and even Descartes occasionally used the iron cross ✠ in his *Geometria*, though that may have been simply added by the printers who searched their typeface cabinets for the closest symbol they could find in order to avoid carving a new letter punch. By the end of the sixteenth century it had taken on a variety of forms from ÷ (our division sign) to = (yes, our equal sign) to—and, eventually, after experiments with other promising options, to $\sim$, the symbol Pierre Hérigone used for minus in the 1634 edition of his *Cursus mathematicus*, an elementary mathematics book that has a reputation for "an almost reckless eagerness to introduce an exhaustive set of symbols."[6] The straight horizontal line symbolizing minus that we have today was the simplest, but it led to some confusion because it was also used as a dash in a sentence. The symbol for minus was not standardized before the eighteenth century. Seventeenth-century manuscripts would often have several forms of minus on the same page.[7]

Multiplication had no fixed symbol for years after Oughtred introduced the symbol × in 1631. Harriot used a dot, and Descartes marked it by juxtaposition. We still use all three notations. However, it is not clear who was responsible—writers or printers? Later, Oughtred would use the colon (:) to denote division. The Arab symbol for fraction used a line to divide two quantities, which varied between $a - b$, a/b, and $\frac{a}{b}$. Our current symbol $a \div b$ is a combination of Oughtred's colon and the Arab line symbol for fraction.[8] Even Leibnitz, who would later create some of the most sensible mathematics notations, used the markings $\smile$ for multiplication and $\frown$ for division. I'm very surprised that they didn't catch on. They are clever, because their reflective duality shows division as simply the inverse of multiplication. Handwriting was the problem; one could be confused with the other.

Our modern symbol for infinity ∞ was a sign that the Romans sometimes used to indicate the number 1,000, hence a very large number. (See figures 4.7 and 4.8.) By the end of the sixteenth century, it foolishly competed with Robert Recorde's horizontal lines and Xylander's vertical lines for the best symbol for equality. That poor symbol ∞ was tossed around to represent one thing and another until 1655, when John Wallis used it in his *Arithmetica Infinitorum* to indicate infinity, yet still, it did not catch on until 1713, when James Bernoulli used it in his *Ars Conjectandi*.[9]

By the time Hérigone's six-volume text was complete and published in 1642, algebra was heavily symbolic. Not everyone was happy with the new form of writing mathematics. Things went too far when several geometry texts were printed with almost no verbal explanation. Soon after Hérigone announced his method, the philosopher Thomas Hobbes in 1648 complained about the shift from verbal to symbolic proofs in geometry:

> Symbols are poor unhandsome, though necessary scaffolds of demonstration...though they shorten the writing, yet they do not make the reader understand it sooner than if it were written in words. For the conception of the lines and figures...must proceed from words either spoken or thought upon. So that there is a double labor of the mind, one to reduce your symbols to words, which are also symbols, another to attend to the ideas which they signify. Besides, if you but consider how none of the ancients ever used any of them in their published demonstrations of geometry, nor in their books of arithmetic...you will not, I think, for the future be so much in love with them.[10]

And De Morgan in 1837 wrote:

> As soon as the idea of acquiring symbols and laws of combination, without giving meaning, has become familiar, the student has the notion of what I will call a "symbolic calculus:" which, with certain symbols and certain laws of combination, is "symbolic algebra:" an art, not a science; and an apparently useless art, except as it may afterwards furnish the grammar of a science. The proficient in a symbolic calculus would naturally demand a supply of meaning.[11]

Chapter 18

The Symbol Master

A seemingly modest change of notation may suggest a radical shift
in viewpoint. Any new notation may ask new questions.
—Barry Mazur

Gottfried Leibniz, a man "of middle size and slim figure, with brown hair, and small
but dark and penetrating eyes," was the genius of symbol creation.[1] Alert to the advantages of proper symbols, he worked them, altered them, and tossed them whenever he felt the looming possibility that some poorly devised symbol might someday
unnecessarily complicate mathematical exposition. He had studied Bombelli and
Viète, and foresaw how symbols for polynomials could not possibly continue into
algebra's generalizations at the turn of the seventeenth century. He knew how inconvenient symbols trapped the advancement of algebra in the fifteenth and sixteenth
centuries.

By the last half of the seventeenth century, mathematics manuscripts were aflame
with symbols, largely due to Leibniz, along with Oughtred, Hérigone, Descartes, and
Newton. Textbook writers and lesser-known mathematicians generated hundreds of
new symbols. Symbol creation at that time was in vogue, but with little understanding of the unanticipated messes that would sooner or later corner creative thought.

As we have seen, Descartes borrowed most symbols and tweaked them to im-
provement. Oughtred introduced hundreds of potential symbols without much concern for their merit. Even when some were clearly problematical, he continued to
use them for the sake of unflappable consistency. This was also true of Hèrigone.

Leibniz, on the other hand, made symbols a priority in his attempts at clear writing. He was convinced that excellent notation was key to comprehension in all matters of human thought. "The true method," he wrote, "should further us with an Ariadne's thread, that is to say, with a certain sensible and palpable medium, which will guide the mind as do the lines drawn in geometry and the formulas for operations, which are laid down for the learner in arithmetic."[2]

In Greek mythology, Ariadne is the beautiful daughter of Minos, king of Crete, and Theseus is the young boy sent from Athens to be sacrificed to the Minotaur in the labyrinth. Ariadne, who has fallen for Theseus, gives him a clew of thread to be unwound as he enters the cave. It was the clue to the way out of the cave once the Minotaur is slain. After Theseus successfully kills the Minotaur and exits the cave, he carries Ariadne off to the island of Naxos and abandons her.

The word *clew*, which originally meant "ball of thread," (and still does) became our word *clue*. Apparently, Leibniz used the symbol of Ariadne's thread to convey the thought that the clue to mathematics and its powers of correct reasoning is in the characteristics of its notation. His calculus notation is so perfectly matched to the basic logical operations and processes of the subject that an ordinary student can follow its thread through labyrinths of reasoning and exit, encouraged by assured comprehension.

Leibniz understood symbols, their conceptual powers as well as their limitations. He would spend years experimenting with some—adjusting, rejecting, and corresponding with everyone he knew, consulting with as many of the leading mathematicians of the time who were sympathetic to his fastidiousness. He didn't take easily to Recorde's equal sign, and so often preferred to use a symbol that looks like a staple ⊓ for equality. I suppose it was meant to suggest a bridge between two sides.

By convention, we now say "y is a function of x" and use the notation $y = f(x)$ to indicate that f is a rule that assigns a unique number y to each and every value x. Leibniz introduced a more restrictive notion in 1692 when he wrote about tangents to curves. For him, a function was simply an expression built from the operations of algebra and analysis—for example, $ax + b\sqrt{a^2 - x^2}$ would qualify because it is built from the algebra operations of addition, multiplication, exponentiation, and extraction of roots. The function concept would go through many revisions before 1837, when it settled for Gustave-Peter Lejeune Dirichlet's brilliant definition that we now use everywhere in mathematics: "y is a function of x, if for every value of x there corresponds a unique value of y." Dirichlet's definition would release all restrictions on how the correspondence is carried out. Descartes did not have such a free definition; he had to associate equations with curves and therefore investigate how one variable moved with another as easily as points in space moved with time.

Among the more than two hundred new symbols Leibniz invented are his symbols for the differential and integral calculus. Anyone who has studied calculus has seen the symbol $\frac{dy}{dx}$, the "derivative of y with respect to x." (See appendix A.)

Why is $\frac{dy}{dx}$ such a good symbol? Without questioning the unjustified symbolic manipulation, $\frac{dy}{dx}$ may be thought of as a fraction; one can multiply both sides of an equation such as $\frac{dy}{dx} = x$ by dx to get $dy = x dx$. How convenient. It turns out that those strange little variables dx and dy actually do follow the rules of algebra, synthetically.

Leibniz's symbols dx, dy of the differential calculus and $\int$ of the integral calculus were vastly superior to any symbols used by other mathematicians working in calculus. They made life in the calculus world far easier than it would have been had Newton's or Fermat's symbols survived. Typesetters objected to the three terraces for symbols like $\frac{dy}{dx}$ that disturbed the spacing of lines on a page.[3] We almost got stuck with his alternative, which looked like dy, as if the top of the d were broken and moved left, or as if that broken off piece was supposed to be a superscripted 1. I would have thought that such a ridiculous symbol would have been a typesetter's nightmare. Lucky for us, it didn't stick.[4]

Such typesetting considerations were a strong force in symbol design. Leibniz followed the common practice of using a vinculum when an operation was to be performed on a group of terms; the vinculum extended over those terms in the group that were to be operated on. It too created problems for typesetters, and so Leibniz invented another way that didn't widen spaces between lines on a page. So, to appease the typesetters, and to make the page look more attractive, he introduced the idea of using a pair of parentheses to indicate which terms are meant to be in the group.

Leibniz was so sure of the success of symbol reform, he bragged:

> I say that when this work is completed it will be the last effort of the human mind, and all men will be happy, since they will have a tool that will laud the intellect as the telescope perfects vision.[5]

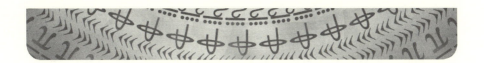

Chapter 19

The Last of the Magicians

Isaac Newton, a man "rather languid in his look and manner, which did not raise any great expectation in those who did not know him," gave figurative credit to those giants on whose shoulders he stood.[1] Popular accounts of Newton recall his famous line, "If I have seen further it is by standing on ye shoulders of Giants," which goes back to the twelfth century when the French Neoplatonist philosopher Bernard of Chartres compared his generation "to [puny] dwarfs perched on the shoulders of giants." Bernard pointed out that we see more and farther than our predecessors, not because we have keener vision or greater height, but because "we are lifted up and borne aloft on their gigantic stature."[2]

Herbert Turnbull, the twentieth century mathematician and Newton historian, tells this delightfully fanciful story about the young Newton:

> In the country near Grantham during a great storm, which occurred about the time of Oliver Cromwell's death, a boy might have been seen amusing himself in a curious fashion. Turning his back to the wind he took a jump, which of course was a long jump. Then he turned his face to the wind and again took a jump, which was not nearly so long as his first. These distances he carefully measured, for this was his way of ascertaining the force of the wind. The boy was Isaac Newton, and he was one day to measure the force, if force it be, that carries a planet in its orbit.[3]

By this time, the telescope was already perfected and the seas around the world explored, yet still, witches were being hanged or burned; traitors and criminals were routinely being beheaded in public squares—their heads parboiled for preservation and hung from posts along busy streets—and alchemy was a hot endeavor for even the brightest of scientists in the face of the new science of chemistry. Even Newton was a strong alchemy experimenter.

Mathematicians of the seventeenth and eighteenth centuries were moderately freed from the classical Greek insistence on mathematical rigor, and empowered by their intuition and speculation about the infinitely large and the infinitely small. With infinity, new rules and new notations had to be developed. Definitions were nebulous, methods hazy, and logical arguments compromised by broken links. "Intuition," Tobias Dantzig wrote, "had too long been held imprisoned by the severe rigor of the Greeks. Now it broke loose, and there were no Euclids to keep its romantic flight in check."[4]

Tools of the infinite and infinitesimal, along with an intuitive grasp of the continuum, were being created and accepted. Imaginary numbers were in the vocabulary. Algebra and its astute use of symbols had prepared mathematics for the calculus revolution. Physics was boosted to a science. And Newton had—in the words of Albert Einstein—"the greatest advance in thought that a single individual was ever privileged to make."[5]

Thanks to the British historian of mathematics Derek Thomas Whiteside and his dedicated editing, we have almost all of Newton's papers. When Tom (as friends called him) first began to study Newton's papers in 1958, they were in a mess. As a graduate student at Cambridge working on a thesis in seventeenth-century mathematics history, Tom began to feel that most existing histories of mathematics were doubtful and casual. As the story goes, he asked a librarian at Cambridge if there were any of Newton's manuscripts available and was quickly given eight boxes to browse. It took Tom twenty-three years to complete his editing of the eight volumes.

From time to time, I browse volume VII, which I seem to share with a mouse that has found something mysteriously delicious in the binding.[6] Just a page at a time, for the seventh volume alone is filled with enough information to spend a lifetime in contemplation.

Newton conceived of unknown variables as quantities flowing along a curve. *Fluents*, as he called them, from the Latin *fluxus* ("fluid"), were very close to the things that we now call *dependent variables*, our *x*'s, but limited by their dependence on time.

Here is how Newton saw them in 1704, almost forty years after he first used them:

> Mathematical quantities I here consider not as consisting of least possible parts, but as described by a continuous motion. Lines are described and by describing generated not through the apposition of parts but through the continuous motion of points; surface-areas are through the motions of lines, solids through the motion of surface-areas, angles through the rotation of sides, times through continuous flux, and the like in other cases. These geneses take place in the reality of physical nature and are daily witnessed in the motion of bodies.[7]

How different this notion is from Leibniz's mathematical quantities. For Leibniz, a curve was fixed, static, and described by its equation, composed of an infinite polygon with infinitesimal sides.[8] Newton, on the other hand, thought of a curve as dynamic, a tracing of a moving particle, where any tangent line pointed in the direction the particle would fly off, if it were not confined to the path. He talked of curves as "flows of points" that represented quantities; yet, for the calculus, they amounted to the same things as Leibniz's static curves.

As time changes, the quantity on the curve flows to a new quantity along the curve. The rate of change of a fluent was the "fluxion of the fluent," a mouthful symbolized as singly dotted forms $\dot{x}, \dot{y}, \dot{z}$—so-called pricked letters that all too rapidly were accepted by the world as standard *fluxional* notation. (See appendix B for details.) Curiously, the higher derivatives were denoted by multiple dots above the variables, so $\dddot{y}$ would stand for the eighth fluxion of the fluent y, which meant the fluxion of the fluxion of ... (eight times) of the fluent y; it was as if the old story of exponential indexing had to be retold before anyone came up with the idea of writing something like $\overset{8}{y}$. Our modern Leibnizian notation for the same thing is $d^8 y$, a far more satisfying representation. Just imagine having to read *ddddddddy* for the eighth differential. There may not have been much need for such a high-order derivative in Leibniz's time, but eventually more complicated terms would inevitably appear: Nightmares such as $12dddddddddxdddddydddddz$ would eventually arise. Fortunately, Leibniz's notation has it as: $12d^8x \cdot d^5y \cdot d^5z$.

A further problem was that, by the restrictive notation, fluxions required a context to clarify the conceptual nature of the independent variable, which was generally the time variable t, but not necessarily. The fluxion of x was understood to be relative to the time variable, and so really nothing more than the velocity of x; in Leibniz's notation, that would be $\frac{dx}{dt}$. In Newton's notation, it would be $\dot{x}$, and in modern language it is: the derivative of x with respect to t.

According to Newton, the fundamental task of calculus was another mouthful of flues: to find the fluxions of given fluents and the fluents of given fluxions. However, Newton had several approaches throughout his life and was also an advocate of infinitesimals.[9]

As an example, take $y - x^2 = 0$; we could substitute $x + \dot{x}o$ for x, and $y + \dot{y}o$ for y. (See appendix B for how Newton found the fluxion of x^n.) The o is the letter "o," which is meant to signify a very, very small quantity, but not zero. In fact, it is meant to be what Newton would call an infinitely small quantity, whatever that was meant to be. With that understanding the equation becomes

$$y + \dot{y}o - (x + \dot{x}o)^2 = 0.$$

And equivalently,

$$y + \dot{y}o - x^2 - 2x\dot{x}o - \dot{x}^2o^2 = 0.$$

Since $y - x^2 = 0$, this last equation becomes $\dot{y}o - 2x\dot{x}o = \dot{x}^2o^2 = 0$. Newton would argue that o is small, but not zero, and therefore division by o is perfectly valid. Dividing by o, the last equation becomes $\dot{y} - 2x\dot{x} - \dot{x}^2o = 0$. Newton would now argue something that could be questionable: because the o is meant to be an infinitely small quantity, the terms multiplied by o must be insignificant compared to the ones that are not multiplied by o. Therefore, he could drop the term $\dot{x}^2o$, so the last equation becomes $\dot{y} - 2x\dot{x} = 0$.[10] Dividing by o when o is not zero is fine, perfectly valid. However, when it came to arguing that o is not zero, but infinitesimal (whatever that means), it left a few great questions to hang in the air for the philosopher George Berkeley, the Anglican bishop of Cloyne in County Cork, Ireland:

> It must, indeed, be acknowledged, that he used Fluxions, like the Scaffold of a building, as things to be laid aside or got rid of, as soon as finite Lines were found proportional to them. But then these finite Exponents are found by the help of Fluxions. ... And what are these same evanescent Increments? They are neither finite Quantities nor Quantities infinitely small, nor yet nothing. May we not call them the Ghosts of departed Quantities?[11]

Newton, and for that matter Leibniz, too, wanted it both ways, to have something to call infinitesimal: something that was not zero to divide by, and yet sort of zero to ignore—"ghosts of departed quantities."

In Berkeley's view, Newton's calculus failed to conform to intuitive notions of continuity. The subtitle of his essay alone gives his point of view: *Or a Discourse Addressed to an Infidel Mathematician. Wherein It Is Examined Whether the Object, Principles, and Inferences of the Modern Analysis* [meaning calculus] *Are More Distinctly Conceived, or More Evidently Deduced, than Religious Mysteries and Points of Faith.*[12]

The real argument was over the justification of Newton's ambiguous meaning of limits of ratios where both numerator and denominator tend toward zero, a fine notion that ignored the appreciation of the subtle nuances and difficulties of infinity and continuity. Newton was not thinking of those ratios as true ratios, but rather as limits, just as we think of them today. To Berkeley, it seemed that Newton was dividing zero by zero, meaningless nonsense.

The bishop's complaint was fair; intuition is fine for people like Euler, Fermat, Newton, and Leibniz, mathematicians with good intuition. The danger was that something slyly anarchic could slip through the front gates of calculus disguised as the legitimate heir to a proven theorem. By the end of the eighteenth century, practical applications of calculus and coordinate geometry were exploding, improving human lives and knowledge of the real world without regard to the inconsistencies sneaking through the gates of reason. The invention of calculus advanced architecture, astronomy, artillery, carpentry, cartography, celestial mechanics, chemistry, civil engineering, clock design, hydrodynamics, hydrostatics, magnetism, materials science, music, navigation, optics, pneumatics, ship construction, and thermodynamics—and this list is by no means exhaustive.

By the time of Newton's death in 1727, eyeglasses and newspapers were readily available and affordable. Enormous political changes had enveloped Europe; small ducal states of central Europe had begun condensing through wars and mergers to become kingdoms, while neighbors shaved large regions from Poland and the Ottoman Empire. Populations of cities remained small—London had fewer than 600,000, Paris fewer than 700,000—and wolves still roamed freely outside the cities. Brightly lit coffeehouses and luxurious surroundings were everywhere in the big cities of Europe as well as in university towns, where newspapers were sold each afternoon, and streets were lit at night so people could walk about, discussing politics, philosophy, and the latest scientific discoveries. Europe was seeing a fresh style of life. Coffeehouses were not just places of gossip and news, but places where students and faculty could talk about the books they read, discuss poetry and plays, collect mail, or hear the latest scientific reports. Scientific academies and societies were established with funds for publishing periodicals and money for developing research tools and costly measuring instruments.

In the fifty years following Newton's death, Denis Diderot would complete seventeen large volumes of the first encyclopedia, Edward Gibbon would shock the world with his *Decline and Fall of the Roman Empire*, Jean-Jacques Rousseau would write *The Social Contract*, James Watt would build the steam engine, Mozart would have written serenades and symphonies, Bach would die, and Beethoven would be born.

Though slave trading increased and wars involving countries all over Europe continued over colonies, trade, and sea power, science, art, literature, and practical inventions were about to explode in the Age of Enlightenment. A middle class was becoming informed and beginning to think, not only about politics, but also about science and literature.

Global information highways were in place to spread news of calamity, intellectual fashions, and scientific discovery. The motions of human culture were growing dramatically more sophisticated and would soon lead to greater discoveries, but the motions of the planet, not to mention cannonballs and arrows, seemed to have been essentially determined by calculus.[13] It was an era witnessing steamy horizons of science, when textbook authors were searching for new ways to express mathematics to an increasing population of university students.

Part 3

The Power of Symbols

Curious readers would like to know the deeper secrets lying beneath the sudden explosion of symbol use and the metamorphosis that brought symbols to us in the forms that they now have. There are those special moments that seem obvious to us now, but far-reaching to a thinker in the past.

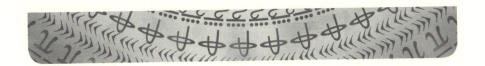

Chapter 20

Rendezvous in the Mind

By relieving the brain of all unnecessary work, a good notation sets
it free to concentrate on more advanced problems...
—A. N. Whitehead

Before the sixteenth century, almost anyone with enough determination could comprehend the elements of almost any mathematical writing. With quill and parchment, a quiet room, an open window with refreshing breezes, enough tallow to keep candles burning through the night, and an inordinate amount of mind-contorting labor, it was still possible to write mathematics in natural language words. Mathematics was readable to anyone who wished to parse its language, its springs, its gears, and its logic.

"Jabberwocky," the *Through the Looking Glass* verse that begins "Twas bryllyg, and the slythy toves" gives an impression of what sensible language sounds like to the uninitiated. It is close to what the infant hears at the stage of trying to make sense of the sounds around him or her. Hear it in connection with "Did gyre and gymble in ye wabe / All mimsy were ye borogoves; / And ye mome raths outgrabe," and something faintly more sensible might come. The Jabberwocky is what we get when we first encounter mathematics—or anything—we don't understand.

By the eighteenth century, the language of mathematics was far too symbolized for people to read without a great deal of preliminary tutoring. It was not so much that the quantity of symbols had grown. Quantity was not the problem; rather, it was that the novice had to learn a new visual language while trying to comprehend new material. Understanding such a language either took a very special expertise or enormously intense work persistence. The language was visual, but the meanings were concealed. Symbols packed in notational sentences provided packets of information whose contents were known only to those who had the time and talent, or the patience to open those packets.

"It's Greek to me," is the colloquial retort we often hear when something is not understood. Greek is not a particularly difficult language. Greek babies learn it as easily as American babies learn English. So why is Greek singled out as an expression of incomprehension? Most likely it's because Greek is not written in those Latin letters that are so familiar to westerners. That unfamiliarity with Greek letters associates itself with a self-commending ignorance.

Mathematical symbols are meant to help us understand. They are meant to help us follow a mathematical demonstration, to make things easy, to charitably give us simplification so that we may have more or less pictorial presentations of what is going on as we read mathematics. But it is true that, like the technical terms of any profession that are beyond our understanding, they become frustratingly Greek—in part because they often *are* Greek.

Alfred North Whitehead dares us:

> If anyone doubts the utility of symbols, let him write out in full, without any symbols whatever, the whole meaning of the following equations which represent some of the fundamental laws of algebra:
>
> $$x + y = y + x$$
> $$(x + y) + z = x + (y + z)$$
> $$x \times y = y \times x$$
> $$(x \times y) \times z = x \times (y \times z)$$
> $$x \times (y + z) = (x \times y) + (x \times z)$$
>
> … The example shows that by the aid of symbolism, we can make the transitions in reasoning almost mechanically by the eye, which otherwise would call into play the higher faculties of the brain.[1]

Although Thomas Hobbes called symbols "necessary scaffolds of demonstration," he also wrote that "they do not make the reader understand [the symbolized] any sooner than if it were written in words."[2]

> For the conception of the lines and figures... must proceed from words either spoken or thought upon. So that there is a double labour of the mind, one to reduce your symbols to words, which are also symbols, another to attend to the ideas which they signify.[3]

In natural language, even the most carefully chosen words drag along concealed meanings that have the power to manipulate reasoning. We learn some words from dictionaries, which give meanings through words we already know or through words we can look up, in turn. We learned (and learn) other words mostly through adjustments of vague meanings, judging how well competing meanings fit each context of use. What is the difference between a chair and a stool, a cup and a mug, a door and a portal?

Symbols of mathematics too sometimes have concealed meanings, but their purpose is to bring along pure thought. It is possible to learn what a mathematical symbol stands for by context. We learn the meanings of mathematical symbols mostly from their definitions: *Mostly*, because in formal mathematics not everyone easily grasps definitions that are not linked to the familiar properties of experience. In a landmark paper, David Tall of the University of Warwick and Schlomo Vinner of Hebrew University point out that many concepts of pending definition are already in the mind in some cognitive structure of personal images before any formal definition is evoked.[4]

Symbolic language surely promotes its own concealed meanings that come from imaginative glimpses into the subconscious, but the best symbols are those that pinpoint meaning and yet permit the mind to quickly roam its databank of similar contextual patterns to compare, to transmit, and to creatively link what is unknown with what is known.

Mathematics uses symbols to express its content with precision. In his notable classic *On Growth and Form*, D'Arcy Thompson asked how we could tell the difference between the shapes of a rainbow and an arc of water shot from a hose.[5] They may look the same; both may even have all the colors of the rainbow. Both are made from droplets of water. In ordinary language, you might say they are smoothly curved arcs that look similar.

But looking at those curves through the lens of symbols, they are very different shapes. The coordinates (x, y) of a point on the rainbow must fit the equation $y = \sqrt{a^2 - x^2}$, whereas the coordinates (x, y) of a point on the arc of water must fit the equation $y = ax^2 + bx + c$, where a, b, and c are fixed numbers that determine the height and width of the curves between their endpoints. One is a semicircle, the other a parabola. No matter how much you tinker with the parameters a, b, and c, the two curves can never be superimposed on each other to become one curve.[6]

With the right and proper symbols, we focus on the patterns, the symmetries, the similarities, the differences that might appear rather dim and blurred through the lens of natural language.

Take the equation $x^2 + y^2 = xy + 4$. Hmm... if only that term xy were not there, we would have a circle of radius 2, which is governed by the simple equation $x^2 + y^2 = 4$. But the term xy is there, so how does it change the circle? It entwines the two variables x and y in such a way that they cannot be separated without some transformation to simplify the equation. However, the symmetry of x and y in the original equation gives a clue to the geometry of the curve. If you swap x and y, you get exactly the same equation. Aha! That can only mean that the curve is symmetric with respect to the line $y = x$. Indeed, by rotating the axes clockwise through a 45-degree angle, and labeling the new axes s and t, the equation magically becomes $3s^2 + t^2 = 8$. This new form has no st term; s and t are not entwined together by multiplication. Graph this nice equation in s and t coordinates, and the picture is an ellipse centered at $(0, 0)$ and symmetric with respect to the s and t axes.

Just as the symmetric form of the equation $x^2 + y^2 = r^2$ cries out CIRCLE! CIRCLE!, so too does the term xy, a multiplicative adhesion of x to y, immediately scream to the left hemisphere of the cerebral cortex, ROTATION! ROTATION! That 45-degree rotation disentangles the variables x and y by making the xy term disappear.

Symmetry in an equation always means some kind of symmetry in the geometry of the curve described by that equation. And so it is for our equation $x^2 + y^2 = xy + 4$. The curve is an ellipse that is symmetric with the two diagonal lines that make a 45-degree angle with the horizontal.

The mind might not have been quick enough to say that the curve is an ellipse, but the equation was quite fast in telling us that, whatever it was, it had to have symmetry about the line $y = x$, because a swapping of x with y could not have changed the curve. All it could have done was to change the names of the variables.

The wires connecting algebra with geometry are unbreakable, yet almost invisible. They make the algebraic process visual. They give us the patterns, associations, similarities, and strange rendezvous in the mind that are veiled by words alone. Accept Whitehead's dare and try to see the geometry of $x^2 + y^2 = xy + 4$ by writing out in full, without any symbols whatever, the equation's whole meaning. It can be done, but not without steaming the blood rushes in the brain.

Symmetry takes many forms. That old quipping question that asks for "the color of George Washington's white horse" is really inviting us to examine the question itself for the answer. When we ask for the square of the number whose square is 4 we are sort of posing a reflective self-answering question. Symbolically, the question is the answer and the answer the question: $(\sqrt{4})^2 = 4$. On the surface, this tautological identity does not ask for new information, nor does it ask for any information. But when we see it symbolically generalized for all positive numbers, as $(\sqrt{x})^2 = x$, our creative talents are stimulated to similar questions: is the identity $(\sqrt[3]{x})^3 = x$ true? What about $(\sqrt[4]{x})^4 = x$? And what about $(\sqrt[n]{x})^n = x$, for any positive integer n?

From this, our talents might leap to a new understanding of the symbol $\sqrt[n]{x}$. If x^n stands for x multiplied by itself n times, and if $(x^n)^m = x^{n \times m}$ for n and m positive integers, then wouldn't it make sense to let the symbol $x^{\frac{1}{n}}$ represent $\sqrt[n]{x}$, "the number whose nth power is x," assuming that such a thing is actually a number. In that way, the algebra confirms what we already know, but it also extends itself to include an arithmetic of exponents. We would have

$$\left(x^{\frac{1}{n}}\right)^n = x^{\frac{1}{n} \times n} = x^1 = x.$$

From there, it is a short hop to knowing that $x^{\frac{m}{n}}$ should stand for $(\sqrt[n]{x})^m$, when n and m are positive whole numbers, and a short leap to a definition that extends n and m to all whole numbers. And from here, we see how symbolism and definition build from one smart idea to the next, using definition, reason, and pattern as guides to more powerful generalities.

This wonderful symbolic form of creating powers and extracting roots provides a grammar for the arithmetic of exponents. Taking powers and extracting roots are inverse operations because they are defined that way. Addition and subtraction are inverse operations; add a number and subtract that same number and you are back where you started. The same works for multiplication and division. In general, a mathematical operation is most useful if there is an inverse procedure that reverses that operation. Such inverse operations are critical to solving equations. For example, to solve the equation $x + 2 = 4$, we subtract 2 from both sides to get $x = 2$. To solve the equation $x^2 = 4$, we take the square root of both sides to get $x = \pm 2$.

Numbers have advanced far from those early beginnings of counting that gave us ten fingers, two eyes, and one nose. They no longer refer exclusively to the things we see or to the things we need to count. Modern mathematics is interested in definition, reason, and patterns that appear to us symbolically. The definitions can even contradict everyday words and intuitive concepts, especially intuitive physical concepts, as long as mathematical rules and symbolic grammar are obeyed. They open the gates to logical worlds that are external to visible nature. Nowhere is this more obvious than when you begin to conceive what lies beyond the rational numbers. Only pure mathematical language, with its highly developed symbolic sense, can see what lies beyond.

At one time, the notion that there could be a number whose square is negative seemed to be beyond the beyond. What could be the use of such an imaginary thing as $\sqrt{-1}$? Use the correct symbolic grammar to solve the equation $x^2 - 2x - 2 = 0$, and you find two reasonable solutions, $1 + \sqrt{3}$ and $1 - \sqrt{3}$. But what pops out when you try that same symbolic grammar on the quadratic equation $x^2 - 2x + 2 = 0$? Two strange solutions, $1 + \sqrt{-1}$ and $1 - \sqrt{-1}$. Take any one of these solutions, square it and subtract it from twice itself and add 2. The result is zero. Separately, these solutions may seem useless, but add them together and you get, simply, 2. In other words, the strange term $\sqrt{-1}$ is annihilated in the process of substituting it in the equation.[7]

If you were just coming from the early sixteenth century reading this chapter, you might be thinking—as you should be—that there is already something suspicious about adding $1+\sqrt{-1}$ to $1-\sqrt{-1}$ to get 2. Such a sum implies that $\sqrt{-1}-\sqrt{-1}=0$. Is that true? A modern answer would be: "Sure, $x-x$ must equal zero, no matter what x might be." Surely that is true for numbers that obey the usual rules of arithmetic. But so far, all we know is that $\sqrt{-1}$ is just a symbol for something that came about as a result of symbolic algebra performed on a quadratic equation. We don't really know anything about $\sqrt{-1}$ other than the fact that it stands for a mysterious *something* with a definitive property that says, "if you multiply it by itself," whatever that could mean, "you get the negative number -1."

You—stranger from the past—might be thinking that $1 + \sqrt{-1}$ and $1 - \sqrt{-1}$ are nonsense solutions because there is no other obvious indication that there are real-world phenomena leading back to a quadratic equation such as $x^2 - 2x + 2 = 0$ (in present-day notation). If you came from the end of the sixteenth century and knew something about graphing quadratic equations in a rectangular coordinate system, you might say that the graph is a parabola whose lowest point is at $(0, 2)$, a point that is two units above the x-axis. There is no x for which the value of y is zero.

But look beyond. Scrap rectangular coordinates and consider something different. What if we now admit to our number system all numbers of the form $a + b\sqrt{-1}$, where a and b can be any numbers that were already admitted to the club of real numbers? You may think that such an admission is silly; however, symbolically, whatever these things are, they act perfectly well within the grammar and syntax of our ordinary numbers. They seem to obey all the laws of ordinary numbers: add two, subtract two, multiply two, divide two, and you get another of the form $a + b\sqrt{-1}$. Believe that all the usual laws apply, and do it! But why is it that—unlike normal numbers such as 1, 2, 3,..., as well as slightly stranger numbers 3/4, π, or $\sqrt{2}$—you have no clear image of what this number represents other than the image of the symbol $\sqrt{-1}$ that you are starting to get used to.

You may think that $\sqrt{-1}$ is already a symbol that represents the square root of minus one, but it was not constructed for that purpose. It emerged from the consequence of algebraic manipulations in trying to solve equations. It may seem as if the negative number -1 got caught under the square root sign accidentally in the process. But, with a bit of arithmetical sleight of hand to avoid details, notice that any number of the form $a + \sqrt{-c}$, with a and c real numbers, can be written as $a + b\sqrt{-1}$, with a and b real numbers. So $\sqrt{-1}$ takes on a virtual importance, and hence warrants a special symbol. We denote it by the letter i, inspired by the word "imaginary." Forms represented as bi, where b represents a real number, are called "imaginary numbers," and forms represented as $a + bi$, where a and b are real numbers, are called "complex numbers," complex in the sense of being a mixture of real and imaginary. (Unfortunately, both words—"imaginary" and "complex"—are solidly anchored in the mathematical vocabulary. *Unfortunately*, because they are the names of classes of numbers that are neither imaginary nor complex.)

It may come as a surprise that the symbol i (even though it is just an abbreviation of the word "imaginary") has a marked advantage over $\sqrt{-1}$. In reading mathematics, the difference between $a + b\sqrt{-1}$ and $a + bi$ is the difference between eating a strawberry while holding your nose, missing the luscious taste, and eating a strawberry while breathing normally.

Numbers? Why are we calling these things numbers? We once thought a number to be a *count* of something—fingers, toes, sheep, days, drachma, eyes, ears, and noses. Then we thought a number was a *measure* of something, which could be fractional or even irrational. But what do these so-called complex numbers count or measure? Perhaps we should call them "pairs of numbers," but even that would not satisfy our usual sense of what a number is. They are not even pairs of numbers, for there is that sticky thing attached to the second number of the pair.[8]

We have images of the integers as a line. Each integer positioned as if it were measuring a unit distance from 0, positive integers to the right, negative to the left. The same goes for rational numbers and real numbers that can be written as possibly never-ending decimals. There is something in the cumulative consciousness of civilization that begs for a mental picture of numbers, even if that picture is fuzzy. But to visualize complex numbers requires something more inventive. (See appendix D.)

At one time, the concept of number was represented as a simple adjective: "ten" fingers. Much, much later it became a noun: "ten," without regard to specific unit-nouns. But ever since the mid-sixteenth century, when symbols flooded into the language of mathematics, the definition of number has conceptually broadened so as to include an act or a mode of being. We now have i, a number that is an action: the act of a rotation of 90 degrees.

About those complex numbers: Way back when Cardano's formula for solutions to cubic polynomials led to the idea that imaginary numbers—in spite of their regrettable name—might be useful, even the most respected mathematicians and philosophers were bewildered by their mysteries. Partly to blame was their apparent inapplicability. If roses by any other name would smell as sweet, then $\sqrt{-1}$ called anything other than "imaginary" would be just as real. It is an unfortunate name, imaginary or not.

Negative numbers came from playing with equations such as $x + a = b$. Imaginary numbers came from playing with equations such as $x^2 + a = b$. So the sense or nonsense of the equation rests on the relationship of the symbols a and b and the legitimacy of square roots of negative numbers. When a is greater than b, there is a problem: a number times itself would end up negative. Such would be nonsense, unless we admit square roots of negative numbers to the number club. To make sense of this nonsense, we must revisit the fundamental question of mathematics—what is number?—so that the solutions to all equations of the form $x^2 + a = b$ have meaning. We want $\sqrt{b-a}$ to have meaning—*always*—as long as a and b are rational numbers.

Perhaps, in some mysterious way, there is some legitimacy to this nonsensical symbol i, something more real than imaginary. Perhaps those strange, meaningless symbols can be used in some way to lead to solutions of problems and yield valid results.

Whitehead once quipped:

> A symbol which has not been properly defined is not a symbol at all. It is merely a blot of ink on paper which has an easily recognized shape. Nothing can be proved by a succession of blot, except the existence of a bad pen or a careless writer.[9]

A complex number $x + iy$ turns out to be just a pair of real numbers (x, y) that must obey a list of easy rules and is spectacularly useful for solving problems about fluid flow, heat conduction, gravity, and almost the whole of mathematical physics. The pictorial representation of the rules for adding and multiplying pairs of complex numbers is surprisingly easy, and meaning for such operations are also surprisingly simple.

One of the wonderful things about mathematics is that—by its best symbols—its progression expands its vision. Multiply any real number by –1, and you have made every positive number negative and every negative number positive. Looking at the real number line graphically, you have spun the whole number line 180 degrees from its original display. Numbers that were growing toward the right become numbers that are growing toward the left. Multiply any complex number by i, and you have rotated it counterclockwise 90 degrees in the two-dimensional plane.

When you try to construct a three-dimensional number system based on triples (x, y, z), you inevitably end up with a system of numbers with nasty things called "zero divisors" (nonzero numbers whose products are zero) that mess up the normal algebra used to solve equations. So skip three-dimensional space and go to four, the next dimension where it is possible to form a number system obeying the associative law—where $a \cdot (b \cdot c) = (a \cdot b) \cdot c$—that has no zero divisors. There is a price, of course: we must give up the commutative law—$a \cdot b$ is no longer equal to $b \cdot a$, as it was for all the numbers we've encountered so far.

The "quaternions," as the nineteenth-century Irish mathematician William Rowand Hamilton called them, belong to a new number system in four dimensions that contain the complex numbers and a multiplication system that obey all the laws of algebra, except the commutative law. Hamilton discovered them on a walk in Dublin with his wife. "I then and there felt the galvanic circuit of thought close," he wrote, "and the sparks which fell from it were the fundamental equations between i, j, k; exactly such as I have used them ever since."[10] (See appendix E for more on quaternions.) Reconsider Whitehead's dare: try writing out the whole meaning of the fundamental equations of the quaternions without any symbols whatever.

Chapter 21

The Good Symbol

The first appearance of the symbol π came in 1706. William Jones (how many of us have ever heard of him?) used the Greek letter π to denote the ratio of the circumference to the diameter of a circle.[1] How simple. "No lengthy introduction prepares the reader for the bringing upon the stage of mathematical history this distinguished visitor from the field of Greek letters. It simply came, unheralded."[2] But for the next thirty years, it was not used again until Euler used it in his correspondence with Stirling.

We could accuse π of not being a real symbol. It is, after all, just the first letter of the word "periphery."[3] True, but like i, it evokes notions that might not surface with symbols carrying too much baggage. Certain questions such as "what is i^i?" might pass our thoughts without a contemplating pause. Pure mathematics asks such questions because it is not just engaged with symbolic definitions and rules, but with how far the boundaries can be pushed by asking questions that everyday words could ignore. You might think that i^i makes no sense, that it's nothing at all, or maybe a complex number. Surprise: it turns out to be a real number![4]

It seems that number has a far broader meaning than it once had when we first started counting sheep in the meadow. We have extended the idea to include collections of conceptual things that include the usual members of the number family that still obey the rules of numerical operations. Like many of the words we use, number has a far broader meaning than it once had.

Ernst Mach mused:

> Think only of the so-called imaginary quantities with which mathematicians long operated, and from which they even obtained important results ere they were in a position to assign to them a perfectly determinate and withal visualizable meaning.[5]

It is not the job of mathematics to stick with earthly relevance. Yet the world seems to eventually pick up on mathematics abstractions and generalizations and apply them to something relevant to Earth's existence. Almost a whole century passed with mathematicians using imaginary exponents while a new concept germinated. And then, from the symbol i that once stood for that one-time peculiar abhorrence $\sqrt{-1}$, there emerged a new notion: that magnitude, direction, rotation may be embodied in the symbol itself. It is as if symbols have some intelligence of their own.

What is good mathematical notation? As it is with most excellent questions, the answer is not so simple. Whatever a symbol is, it must function as a revealer of patterns, a pointer to generalizations. It must have an intelligence of its own, or at least it must support our own intelligence and help us think for ourselves. It must be an indicator of things to come, a signaler of fresh thoughts, a clarifier of puzzling concepts, a help to overcome the mental fatigues of confusion that would otherwise come from rhetoric or shorthand. It must be a guide to our own intelligence. Here is Mach again:

> In algebra we perform, as far as possible, all numerical operations which are identical in form once for all, so that only a remnant of work is left for the individual case. The use of the signs of algebra and analysis, which are merely symbols of operations to be performed, is due to the observation that we can materially disburden the mind in this way and spare its powers for more important and more difficult duties, by imposing all mechanical operations upon the hand.
>
> The student of mathematics often finds it hard to throw off the uncomfortable feeling that his science, in the person of his pencil, surpasses him in intelligence—an impression which the great Euler confessed he often could not get rid of.[6]

A single symbol can tell a whole story.

There was no single moment when x^n was first used to indicate the nth power of x. A half century separated Bombelli's $1.\underset{\smile}{2}$, from Descartes's x^n. It may seem like a clear-cut idea to us, but the idea of symbolically labeling the number of copies of x in the product was a huge step forward. The reader no longer had to count the number of x's, which paused contemplation, interrupted the smoothness of reading, and hindered any broad insights of associations and similarities that could extend ideas. The laws $x^n x^m = x^{n+m}$ and $(x^n)^m = x^{nm}$, where n and m are integers, were almost immediately suggested from the indexing symbol . Not far behind was the idea to let $x^{\frac{1}{2}}$ denote $\sqrt{x}$, inspired by extending the law $x^n x^m = x^{n+m}$ to include fractions, so $x^{\frac{1}{2}} x^{\frac{1}{2}} = x^1$.

Further speculation on what n^x might be would surely have inspired questions such as what x might be for a given y in an equation such as $y = 10^x$. Answer that and we would have a way of performing multiplication by addition. But Napier, the inventor of logarithms, already knew the answer long before mathematics had any symbols at all!

Symbols acquire meanings that they originally didn't have. But symbolic representation has, likewise, the disadvantage that the object represented is very easily lost sight of, and that operations are continued with the symbols to which frequently no object whatever corresponds.

Ernst Mach once again:

> A symbolical representation of a method of calculation has the same significance for a mathematician as a model or a visualisable working hypothesis has for a physicist. The symbol, the model, the hypothesis runs parallel with the thing to be represented. But the parallelism may extend farther, or be extended farther, than was originally intended on the adoption of the symbol. Since the thing represented and the device representing are after all different, what would be concealed in the one is apparent in the other.[7]

Chapter 22

Invisible Gorillas

...Hark! the rushing snow!
The sun-awakened avalanche! whose mass,
Thrice sifted by the storm, had gathered there
Flake after flake, in heaven-defying minds
As thought by thought is piled, till some great truth
Is loosened, and the nations echo round,
Shaken to their roots, as do the mountains now.
—Shelley, *Prometheus Bound*[1]

A frog easily catches insects in motion, but will not bother a most appetizing fat housefly sitting directly in front of him. A fly could safely crawl onto the frog's back without any worry of being gobbled up. Place a plate of dead flies in front of the frog and he will sit there like a stone garden ornament. The poor frog would starve to death rather than attack something that is not moving.

The pond in my yard is filled with frogs of all sizes. I see one, but he does not see me—not really. His eyes don't move, but if his body moves he reorients himself, rotating the world with him. Pluck a long reed of grass from the banks and very slowly move the tip toward the frog's eyes. Keep it steady, and the frog will just sit there, as if staring across the pond. Wiggle the end of the reed, and a tongue will dart from the easily fooled frog to catch the reed. But should an insect pass his unblinking visual field, he will snatch it as quickly as a bullet leaves a .357 Magnum.

"And should he miss?" I once asked Jerry Lettvin, the neurobiologist who wrote the seminal paper on what the frog really sees.[2] "Well," Jerry said, "he will remember that moving thing as long as it stays within his field of vision and he is not distracted."[3]

The frog sees movement. He is able to catch flies so well because there are no other obstructive confusions in his visual field. Humans are okay at catching flies when the background is white or a solid color, but the moment the fly moves into a field with confusing background, we lose track of its movement.

Symbols provide a blank background on which we may contemplate unadulterated meaning. They help us see, as if through frogs' eyes, to distinguish ingredients: the essentials from the disposable, the elementals from the jumbles.

You put the equation $x^2 - ab = 0$ before me, and I immediately know that $x = \pm\sqrt{ab}$. But I would also see a square and a rectangle that are aching to be compared. I see a little poser wiggling in before the whiteboard of my mind: "What is the side of a square that has the same area as the rectangle of length a and width b?"

Every mathematician I know would see the same little poser. It would be like putting the musical notation

before the eyes of a musician or, for that matter, anyone who can read music. The mind would hear the four-note "short-short-short-long" motif played twice, and know it to be the opening motif of Beethoven's Symphony No. 5 in C minor, Op. 67.

My little poser would conjure up several cerebral images. There might be a geometric one, where two figures, a square and rectangle, are compared in such a way that I could reconfigure the rectangle to make it into a square. Since a and b do not have specific values, the exercise can only be one of symbolic manipulation. I would resort to the rules of algebra learned in school: add ab to both sides and extract the root of ab to get $x = \pm\sqrt{ab}$.

Meanwhile, my mind would probably rush though hundreds of specific cases almost at once, searching for connections to all the other times such an equation had been seen, where $a = \cdots - 3, -2, -1, 0, 1, 2, 3, 4, \sqrt{2}, \pi \cdots$, and the same with b. Those particular cases would give me fixed images of specific rectangles. When $a = 3$ and $b = 12$, and the multiplication is performed, I would recognize the perfect square whose area is 36 square units and whose side is 6 units long. But if $a = 3$ and $b = 10$,

I would be looking for a square whose area is 30 square units, at a slight loss over what the side of such a square will be.

At that point, my mind must go into a secondary mode. I must think in terms of extracting roots and all the information I have accumulated over the years about extracting roots. Ugh! What is the square root of 30? It's a tough one, if I have not recently made the calculation to find it. I would recall that it is less than 5.5 and more than, say, 5.2. But then I would tell myself that I'm not really interested in what it is exactly, and that $\sqrt{30}$ is either good enough or that it is $\sqrt{2 \times 3 \times 5}$, which is also good enough.

In the early nineteenth century, the German naturalist Gotthilf von Schubert wrote the most influential book on dreams, a book that is reputed to have influenced Freud and Jung. Von Schubert observed that we dream in a *traumbildsprache* ("dream visual language"), "a higher kind of algebra," not in a verbal language. The pictures we see are the symbols of myths and rituals of peoples throughout the world. But the pictures are mostly silent. Except for the few occasions when there is verbal activity, the sounds spoken by the dreamer appear as muffled garbling to anyone who is awake and listening. Even nightmare screams are silent in dreams; the distressed dreamer struggles to get out the faintest noise.

In the 1940s, the American psychologists Calvin Hall and Vernon Nordby began collecting dreams. Over the next thirty years, they compiled the extracts of more than 50,000 dreams from people of all ages and from all around the world. By a classification scheme, they discovered that dreams of random groups scattered around the world are more similar than different.

Why do the themes of dreams recur in so many different cultures around the world? Hall and Nordby called them "typical dreams"—"These *typical dreams,* as we shall call them, are experienced by virtually every dreamer. These typical dreams express the shared concerns, preoccupations and interests of all dreamers. They may be said to constitute the universal constants of the human psyche."[4]

Why? The likely answer is that picture language predates verbal language, and that dreams are a part of the collective unconscious—Jung's theory. Pictures in the mind once gave humans a powerful protolanguage for survival. There was a time when humans *thought* and *communicated* with noises no more sophisticated than a hound's yelp. The first verbal language was likely grunting or need-indicators vocalized by a single vowel. The bird does not *think* by whistling to itself. It goes on about its business of building nests by model images of the representative nest, without any sure idea of what it is doing, and yet it builds its nest from the instructions of its instinctive perception coming from its central nervous system. It goes about its daily business with a species-specific sense of behavior patterns.

The evolution of our capacity to make sense of visual meanings happened long before we had any communicating tool that we would now call language. So we should be inclined to expect that images be more at the core of intuitive cognition than words. We may have silent chats with ourselves, have dialogues with the self, but the images we see are more primal, and don't require words for us to understand what we see. We may verbally translate our images, but such translations are not necessary for thought.

Images and sounds make the invisible sensory expressions of thought visible and the inaudible sensory expressions of thought audible. For sensory thought to be at all useful, there must be some transformational code that brings some image or sound into consciousness. Images are primal. Written words and mathematical symbols are invented. A walk in the woods takes in a great many images—the random stones, the fallen branches, the wet leaves at the edge of the trickling stream, the green grass, the blue sky peeping through the treetops. These are not verbalized. Rather they become images stored into the gazing thoughts compartment who-knows-where in the brain. They get confused and synthesized with other experiences through comparison and association with similar memories of real events and mental images.

It may be true that symbols in mathematics are distinct from symbols that come from experiential senses that are apparent in dreams, myths, rituals, and poetry, as the American philosopher Suzanne Langer suggested in her 1967 final and seminal work, *An Essay on Human Feeling*. However, the moment we read an equation—simple or complicated—images form in the mind along with verbal reflections that suggest multiple metaphorical connections and associations with what had been seen before.[5] One might say that a person's knowledge is no more than the cerebral recollections of images and verbal reflections. Like a walk in the woods, a synthesis is born from all the past symbolic explorations of the collective mathematics journey, a synthesis that issues in a process of abstraction. Here is what Langer wrote back in 1954:

> [The power of understanding symbols] issues in an unconscious, spontaneous process of abstraction, which goes on all the time in the human mind: a process of recognizing the concept in any configuration given to experience, and forming a conception accordingly.[6]

Unlike visual conception, verbal reflection needs a wee bit of conscious help to overcome its impermanence so it can form meaning and be safely stored in long-term memory. Again, here is Suzanne Langer:

> No assignment of meaning is conventional, none is permanent beyond the sound that passes; yet the brief association was a flash of understanding. The lasting effect is, like the first effect of speech on the development of the mind, to make things conceivable rather than to store up propositions.[7]

Our unconscious involuntary thoughts interact with our conscious thoughts to give meaning to our thinking. How could such meaning come, without indescribable perceptions unconsciously suggested by sense experiences of the real world? Although symbols and words help to form our thoughts and viewpoints, only symbols can shape the complexities of communicable ideas into cohesive expressions. Of course, words can do the same and are necessary to explain thoughts and ideas. But because words must fleetingly deal with one thought at a time, they can quickly fall into the cracks of confusion in the onslaught of oncoming words that are necessary to complete the thought. Though symbols in mathematics are tightly defined by the explaining words that define them, they awake suggestive thoughts that would not be directly intended by the words themselves.

When it comes to algebra, visual conception is beyond any similarities in the physical world. That's okay; as we've noted, it's not the job of mathematics to be concerned with the physical world, nor with what we call "reality." Symbolic consistency and meaning are essentials of mathematics. So is certainty. So is imagination. So is the creative process. So is hypothesis. So is belief beyond experience. So is adventure of knowledge. And, in today's complexity, there is no better way to do the job of mathematics than by symbolic envisagement.

These days, mathematical representations come in all types. Some are iconic, in that they resemble what they represent. Some are truly symbolic. And some are used purely for indexical purposes. In her book *Representation and Productive Ambiguity in Mathematics and the Sciences*, Emily Grosholz contends, "Which representations we have at our disposal and how we combine them determines how we can formulate and solve problems, discern items and articulate procedures, supply evidence in arguments and offer explanations. And how the representations should be understood, their import and meaning, must be referred to their use in a given tradition of problem solving."[8]

At a recent math conference in Boston, I designed a little experiment in symbolic cognition that involved interviews of several colleagues, all professors of mathematics. It was hardly an acceptable scientific design. At the center of my laptop screen was a symbolic expression involving a square root and a few squares. The specific expression is not important. (See appendix C for a transcript of an interview.) Each interview started with me pointing to the screen of my laptop, asking: "What goes through your mind when you see such a thing as this?" In each case, there was a long pause, after which I would tell the subject that there is no right or wrong answer. Then, there would be a stab at an answer, usually some geometric argument that had to do with the graph of the equation. "This might have something to do with an ellipse," was one attempt. "It's a cone," was another.

At one point, an obvious hint in the form of a new expression would fade in at the top of the screen with two arrows pointing directly to the original equation. It would remain displayed for a full 10 seconds. The subjects were looking directly at the screen when, after 10 seconds, the equation and arrows faded out.

I interviewed nine people this way; all but two tried to tie the question to the graph of the equation in question. But after the strange 10-second display of the fading-in-and-out expression, two of my subjects got the same idea.[9] Their answers were exactly what the hint had suggested, the solution to a general quadratic equation. There was no verbal indication that either of the two subjects was aware of the fade-in/fade-out equation on the screen. At the end, I asked each subject if anything unusual was seen on the screen of my laptop while they were contemplating the question. Their eyes widened. Everyone, including the exceptional two, claimed to have seen nothing fade in or out.

What would have happened had my laptop displayed the equation rhetorically instead of symbolically? Inevitably, anyone would have translated the wording into symbols. But if we were still living in Gerolamo Cardano's mid-sixteenth-century mathematics world, a world that would have known the solution to a quadratic equation (as Brahmagupta had even as far back as the seventh century) expressed only in words, would that association have come so quickly from a verbal description of the hint that came from my fade-in-and-out equation?

Asking a question such as "What goes through your mind when...?" the way I did reminds us of how social science experiments were performed back in the mid-twentieth century, when there were few mechanisms in place for measuring responses. My sample size was so small, there was no real way to tally the frequencies of the answers. Moreover, even if the sample size were far larger, the experiment would have to take into account two of the modern notions of how recent associations are yoked to immediate cerebral responses. We now know that we are all subject to both the "priming effect" and the "anchoring effect," two well-studied subconscious supremacies that can manipulate our conscious reasoning.

The priming effect tells us that our actions and emotions are affected by our experience of recent events. For instance, if you were asked to fill in the blanks of the word "S_ _P," you would likely write "SOAP" if you had just washed your hands, and that you would likely write "SOUP" if you had just sat down for dinner. There are also Freudian symbolic connections that claim that you would write "SOAP" after being asked to think of an action that brings you shame, Freudian in the sense that the soap is a cleanser for the stained soul.[10]

The anchoring effect is different. It unconsciously locks us into a small range of associative thoughts with a tendency to anchor our opinions to some immediate bias. An experiment conducted by Tversky and Kahneman back in 1974 asked subjects to guess the percentage of African countries listed as members of the United Nations. A wheel marked with numbers from 0 to 100 was spun. The wheel would come to rest at a number—say, X. The subjects were asked to first indicate whether X was higher or lower than the answer to the question. Following that, the subjects were asked to estimate the value of the quantity by moving upward or downward from that number. The bizarre outcome was that, for the group that saw the wheel land on 10, the median estimate of the percentage of African countries that were members of the United Nations was 25, and, for the group that saw the wheel land on 65, the median estimate was 45. The actual answer is 30. What could a wheel of fortune have to do with the number of countries belonging to the United Nations in 1974?

In asking a question such as "What goes through your mind when …?" as an experiment, we should understand that anchoring and priming will play significant roles in how responses will be biased by immediate associations. A subject who had just been exposed to one kind of thought may be both primed by it and anchored to it. However, I do believe that when reading mathematical symbols, priming and anchoring may positively lead to new results. In reading mathematical symbols, priming and anchoring work together to constructively guide us through barrages of competing associations that simultaneously beckon for some preferential attention. Anchoring may lock us into immediately preceding thoughts, but that may be a good thing when reading mathematics.

The cognitive psychologists Keith Stanovich and Richard West tell us that we think on two levels, which they label "System 1" and "System 2" so as not to prejudice their experiments.[11] For now, I prefer to call them "auto mode" and "focus mode." Auto mode requires no effort and no sense of conscious control, whereas focus mode means there is a controlled effort to keep the object of thought in focus. We can drive a car along an empty highway listening to music and talking with a child who asks what $2 + 2$ is. That takes no effort. When we read books such as the one you are now reading, we are using both focus mode and auto mode. When we read mathematics, no matter how simple, we are using both. We use both in the sense that the focal can have an effect on the auto. How's that?

Christopher Chabris's and Daniel Simons's now famous "Invisible Gorilla" experiment showed how the focus mode might interfere with the auto mode.[12] Experiments in "inattentional blindness"—the failure to perceive a visible, unexpected object while attention is focused on a task—are not new.[13] The latest are based on auditory studies and their visual analogues that were conducted in the 1950s and 1960s. Chabris's and Simons's Invisible Gorilla experiment is striking. With students as actors, they made a one-minute film of two teams—one in white shirts, the other in black—moving and passing a basketball. Subjects were asked to silently count the number of passes made by the white-shirted players while ignoring any passes by black-shirted players. Immediately after the video ended, the subjects were to report how many passes they had counted. Halfway through the video, a female student in a full-body gorilla suit walked across the court, stopped directly in front of the camera, thumped her chest, and walked off. At the end of the video, the subjects were asked a series of questions:

Q: Did you notice anything unusual while you were doing the counting task?
A: No.
Q: Did you notice anything other than the players?
A: Well, there were some elevators, and S's painted on the wall. I don't know what the S's are for.
Q: Did you notice anyone other than the players?
A: No.

Q: Did you notice a gorilla?

A: A what?![14]

About half of the subjects did not notice the gorilla! A gorilla that walked directly through the center of the court! The gorilla did not contribute to the task; hence, there was a deficiency of attention, and hence, the gorilla was invisible.

The experiment was not intended to tell us how the mind works when we are doing mathematics; rather, it was to show that we may miss the unexpected during calls for concentrated visual attention. Yet, in a limited sense, it does apply to mathematics. I know many mathematicians who, in deep thought over a problem, would not be aware of a live gorilla in the room. How many times have I denied that I ever heard something my wife told me? The gorilla might be in the room and I wouldn't know it, but should she walk across the problem I am working on, she wouldn't need to thump on her chest for me to know she is there.

The Invisible Gorilla experiment applies only to inattentional blindness of unexpected objects in visual fields. What about gorillas *in* mathematical problems? Return to the experiment I performed at the Joint Meetings of the American Mathematics Society and the American Mathematical Association. Two people saw something that reminded them of something they had seen before. What made two subjects see something algebraic and seven subjects search for a graphical connection, when all nine must have subconsciously witnessed the same hint? Perhaps it was simply a matter of mathematical brain type, as Poincaré would have put it. He wrote:

> Among our students...some prefer to treat their problems "by analysis," others "by geometry." The first are incapable of "seeing in space," the others are quickly tired of long calculations and become perplexed.[15]

There are answers to the question of why some people would immediately resort to geometric daydreaming and others to analytic. There are, of course, many works that attempt to understand the psychological and neuropsychological aspects of intuition and the creative talents involved in understanding mathematical proof, techniques, or computation—Hadamard and Poincaré in the early twentieth century, Dubinsky and Polya in the twentieth century, and continuing with George

Lakoff, David Geary, Stanislas Dehaene, David Tall, and others in this century.[16] But experiments in cognitive neuroscience to find answers to the question of what goes through the mind in reading a rhetorical statement as opposed to its symbolic equivalent are just now in their delicate stages of infancy. I'm not talking about some kind of phrenological pinpointing of mathematical thought in the brain, or some GPS of the brain's neurophysiological highways to localize mathematical thinking. My question should not be so tough to answer, and yet it seems to be. It also may not be of much interest to experimental psychologists. The real difficulty with experiments in mathematical cognition is that humans have too many distinct and imaginative thinking schemes to make analysis definitive and interesting. We all think somewhat differently with brains that are exquisitely different, using richly assorted thinking styles that contribute to and account for the preciousness of being human.

The most interestingly close work on such a question comes from Stanislas Dehaene's laboratory, the Cognitive Neuroimaging Unit at the CEA (Commissariat à l'énergie atomique et aux énergies alternatives) in Paris. Dehaene and his students used electroencephalographic techniques—under the idea that the brain activity generates electric current—to study differences in brain activity between contemplations of numerals and words. They flashed Indian numerals and number words on a computer screen to find out how long it takes a brain to decide, with millisecond accuracy, that 4 is smaller than 5. A few surprises followed.

Volunteers were asked to press one key with the left hand for numbers smaller than 5 and another with the right hand for numbers larger than 5. Minute changes in scalp voltage generated by brain activity were recorded on a timeline of milliseconds from sixty-four scalp electrodes.[17] For the first hundred or so milliseconds, electric potential registered close to zero; then, a positive potential registered at the rear of the scalp, suggesting that visual areas of the occipital lobe were activated. At this stage, when visual actions engaged, Dehaene found no perceptible difference between Indian numerals and English number words. But then, suddenly, words such as "four" began to generate negative potentials only in the left hemisphere, while digits such as "4" produced a potential in both hemispheres simultaneously.

An entire single event of mentally processing whether a number is greater or less than 5—from number recognition to the motor response of pressing the selected button—on average takes less than half a second. So what happens in the process? At around 150 milliseconds, a miscellany of specialized areas of the visual cortex become active, presumably in recognizing the shape of the numerical symbol without attributing any meaning. Then, at around 190 milliseconds, when it is assumed that the numerical quantity is first being encoded, a difference in electrical potential amplitude was found between digits that were close to 5 and digits that were far from 5; presumably, digits that were far from 5 are more easily distinguished as being greater than or less than 5.

Now the first surprise: Though number words generated a negative potential only in the left hemisphere, and digit numerals produced currents in both hemispheres, the electrical potential amplitudes were similar for both number words and digit numerals. In other words, the inferior parietal region (a region attendant to language and mathematical operations) seems to recognize the abstract magnitude of numbers without regard to notation.

The second surprise came at the first micro-moment of the motor response, just after the number comparison was completed and the answer was ready—that is, between about 250 milliseconds from the time the digit or number word appeared on the screen to about 330 milliseconds. At that instant, there appeared to be a distinct voltage difference between the right and left premotor and motor areas. When a subject first prepared a right-hand response, the left-hemisphere electrodes indicated a negative potential. Preparation for a left-hand response generated a negative potential in the right hemisphere. This suggests that it takes between a quarter and a third of a second for the mind to recognize the shape of the digit or number word and decipher its quantitative substance.

And then the third surprise: On average, it took another 50 milliseconds for the finger muscles to contract and to actually press a button. Even with a simple task of trying to decide whether a number is less than 5 or greater then 5, people make mistakes. When that happens, there is an immediate and intense negative potential at the frontal lobes—an area associated with controlling actions and inhibiting unwanted behavior—that suggests a mental attempt to correct the error. This, which happens at the astonishingly fast speed of less than 70 milliseconds after pressing the wrong button, is surprising because it shows that the reaction to the unintended response is a psychodynamic one.

In Dehaene's experiments, precise brain activity location was compromised by the skull's tendency to diffuse electric potential. To pinpoint distinct regions where electrical activities happen, more invasive procedures would have been needed, procedures involving electrodes implanted in the cortex itself. Such a procedure can be done only under exceptional circumstances such as with patients suffering from debilitating seizures. Those procedures of intracranial electrode probes were performed at Yale by Truett Allison and Gregory McCarthy in 1994 with results that pinpoint two neighboring regions of the visual processing area of the brain: one reacted to words exclusively, the second only to Indian numerals, and a third only to faces.[18]

Dehaene, Allison, and McCarthy are not suggesting that the brain behaves in some phrenological way. They know that even the simplest functions activate large and distinct cerebral geography. They know that no one area of the brain can think for itself. Although no single region of the brain can perform even the simplest thinking task, there does seem to be, however, some concentration of electrical activity for minute instants of specialized brain activity such as reading a word or performing a calculation. Dehaene thinks of the brain as "a heterogeneous group of dumb agents. Each is unable to accomplish much alone, but as a group they manage to solve a problem by dividing it among themselves."[19]

No doubt, there is a difference between reading a phrase of words and reading a symbolic mathematical phrase. For instance, the phrase "one added to four plus the difference between three and two" is read differently than the phrase $((3-2)+4)+1$. The question of difference splits into two competing questions raised by Dehaene:[20] Does our comprehension of mathematical expressions come from our capacity for processing language structures? Or is it language-independent, relying on some visual system for parsing strings of mathematical symbols?

The answer narrows to whether mathematics is primarily visual-spatial, or predominantly linguistic. There seems to be a difference in patients with severe aphasia or dementia who retain the ability to comprehend simple symbolic algebra problems.[21] Patients with lesions outside the language areas of the brain sometimes have great difficulty conceptualizing Indian numerals and their corresponding word equivalents.[22]

Dehaene specifically investigated micro-ocular behavior of subjects processing the expression $1+(4-(2+3))$ in order to study how the brain represents mathematical expressions with so-called nested embeddings, and how closely related symbolic numerical computations with nested structures are to the parsing of equivalent linguistic expressions.

In reading a sentence of text, we must be aware of punctuation within the sentence before the actual reading. A "What" at the beginning of text does not always indicate a question. Even my Microsoft Word spellchecker insists that the preceding sentence is a question.

Like question marks at the ends of sentences, nesting phrases within a sentence requires the reader's eyes to be slightly ahead of cognition. A sentence from William Faulkner's short story in *Go Down Moses* titled "The Bear" has us hanging till its completion:

> It was of the wilderness, the big woods, bigger and older than any recorded document—of white man fatuous enough to believe he had bought any fragment of it, of Indian ruthless enough to pretend that any fragment of it had been his to convey; bigger that Major de Spain and the scrap he pretended to, knowing better; older than old Thomas Sutpen of whom Major of Spain had had it and who knew better; older even than old Ikkemotubbe, the Chickasaw chief, of whom old Sutpen had had it and who knew better in his turn.[23]

Admittedly, a Faulkner sentence may not be a fair example of an English sentence, as Faulkner often tried to put the whole world into each sentence. But I spared you. The story contains much harder sentences; there is another that is probably the longest sentence (more than 1,600 words!) in American fiction. So indulge me, and take a moment to read the sentence again. Parsing Faulkner's sentence requires reading quite a bit ahead of where the eye wants to go, and yet the sentence makes perfect sense as it is being read.

A mathematical expression such as

$$4\left(3x^2 + 2\left(-3 + \left(2 - 3 + (2x + 1)^2 + 1\right)^2\right) - 4\right)$$

is processed only after the eye has scanned the expression in search of the most inner nested sensible operation—namely, $2x + 1$. The full expression is not generally read from left to right, though it may have been instantaneously scanned in that direction.

Little is known about how people comprehend mathematical expressions. Lately, there have been experiments involving functional magnetic resonance imaging (fMRI) and magneto-encephalography (MEG) as well as electro-encephalography (EEG) to measure the cognitive timeline of processing simple algebra and response in symbol manipulation tasks—that is, to measure the brain location and speed of processing very elementary mathematical expressions. In particular, Jared Danker and John Anderson at Carnegie Mellon used fMRI and MEG imaging to study the parsing speeds of subjects solving linear algebraic equations with three terms and one unknown.[24] They attempted to isolate activity in two regions of the brain: the parietal cortex and the prefrontal cortex. The main function of the parietal cortex, which is located behind the frontal lobes and above the occipital lobes, is to integrate sensual information—in particular, to integrate visually perceived data for sense of space and navigation. The prefrontal cortex (the anterior part of the frontal lobes) lies just in front of the motor and premotor areas, and functions as an orchestrator of complex cognitive behavior, choosing which to act on from conflicting thoughts such as good and bad, same and different.

Using algebraic symbol manipulation tasks, Danker and Anderson examined the two regions of the brain in a two-step cognitive process of equation solving, isolating in time transformations of representations and memory retrieval of mathematical facts. The first involves the parietal cortex when visually received information is processed and an image representation is transformed. The second involves the prefrontal cortex when it receives a transformed representation and brings in the retrieval of mathematical facts that interact with the processed representation of the problem. This two-step process continues in a back and forth sequence. For example, when the stimulus was to solve equations such as $\frac{x}{3} + 2 = 8$, the mathematical knowledge comes in after the prefrontal cortex retrieves the difference $8 - 2 = 6$; the parietal cortex transforms the equation and encodes it into $\frac{x}{3} = 6$; the prefrontal cortex takes that transformation to retrieve the multiplication fact $3 \times 6 = 18$; and finally, the prefrontal cortex transforms the equation and encodes it to $x = 18$. Both brain regions behave differently during each step, yet there is a strong interactional relationship between retrieval and representation in mathematical thinking.[25]

Anthony Jansen, Kim Marriott, and Greg Yelland of Monash University studied how experienced users of mathematics comprehend algebraic expressions.[26] They designed memory tasks to examine the role of mathematical syntax in encoding algebraic expressions, and concluded that experienced users of mathematics had an easier time identifying previously seen syntactically well-formed expressions than ill-formed ones. They found that the encoding of algebraic expressions is based primarily on processes that occur beyond the level of visual processing. For example, the well-formed string $7 - x$ is better recalled than ill-formed strings such as $7(x.$ That may not be as surprising as the fact that in the expression $4 - x^2)y - 7$, the $4 - x^2$ is better recalled than if it were seen in the expression $4 - x^2(y - 7)$.

The Monash team later investigated how experienced users of mathematics parse algebraic expressions by examining the order in which eyes scanned the symbols and measuring the durations of ocular fixations.[27]

In reading text, we tend to pause for a few milliseconds at the end of clauses and sentences. That seems perfectly natural; it conforms to the way we speak: a flow of

nouns, a verb or two along with modifying adjectives, adverbs and perhaps a pro-
noun, all coming from a capsulated constituent thought. We do this not just to help
the breathing in vocal communication, but also to convey the sentence structure as
an arrangement of word forms and their mutual relations within the sentence. We
might not expect the same reading scheme when reading a mathematical expression,
yet the Monash team found that we seem to do something similar when reading
mathematical expressions: symbols at the end of an arithmetic phrase were fixated
upon for significantly longer than symbols at the start or middle of the phrase. This
suggests that we "read" algebraic expressions by their syntax, just as we do when
processing sentences of natural language.

Dehaene's experiment examined measured eye movements of mathematically
trained subjects while they computed nested arithmetical expressions such as $4 +
(3 - (2 + 1))$. The eyes moved to the deepest nest, $(2 + 1)$, and worked their way up
the nesting to complete the computation. In other words, the spotting of the deepest
nest was instantaneous with almost no reading of the expression from left to right,
as there would have been in a text form of the expression. It implies that the subjects
quickly parsed the initial string for syntax, relying on operator and parentheses as
cues before their eyes moved to the successive levels of the syntactic hierarchy, at
each step, recovering digit identities that were needed for calculation.

The subjects were young (19 to 27 years old), having had a general mathematics
education at French universities. The strings always used the four digits 1–4 along
with two plus signs, one minus sign, and two pairs of parentheses. There were four
levels of complexity, labeled 0–3. The highest level was 3, which exhibited a mathematically valid string of terms such as $((3-2)+4)+1$. Level 2 was created by swapping the outer parentheses and shuffling the symbols outside them—for instance,
$)(3 - 2) + 4(+1$ would be one such string. Level 1 was created by swapping the inner parentheses and shuffling all the symbols outside them—that is, $+)3 - 2(4+)1($.
Level 0 was a mathematical nonsense string created by a shuffling of all terms—for
example, $4 - +)3)(+2(1$.

Dehaene's team found that mathematically trained subjects were sensitive to the complexity of mathematical expressions, that complexity effects were localized to a set of cortical regions located outside of classical language areas, that the parsing of mathematical expressions starts early on during visual processing, and that that area relies heavily on the fusiform cortex, an area of the brain involved in identifying words from shape-images, recognition of the visual world, as well as visual identification of well-formed mathematical strings.

It seems that particular notational configurations may help us recognize structure in mathematical expressions and process equations—for instance, inappropriate spacing may lead to confusing spatial configuration with syntactical structure, as $2+3 \times 4$ may lead to a different answer than $2+3 \times 4$. So the early stage of processing a mathematical expression may be more like the early stages of word processing; both first identify the parts of the expression to determine if they are legal or not and then go on to the stage of syntactic parsing.

There was a time when even the most respected philologists believed that thoughts happen only through language.[28] By language, those philologists meant words. One of the great nineteenth-century German philologists Max Müller claimed, "We may feel the dark, but till we have a name for dark and are able to distinguish darkness as what is not light, or light as what is not darkness, we are not in a state of knowledge, we are only in a state of passive stupor."[29] We have come a long way from that kind of understanding. What about animal thought?

Chapter 23

Mental Pictures

Years ago, during some summers on Cape Cod, I would jog down a dirt road, passing a giant German shepherd tethered by chain who would bark one long "grrrrrhoff" as I passed—just one. I would answer with a very quiet "whoof," my "hello" in what I thought might be dog language. After a few days into the season, the dog stopped "grrrrrhoffing," and just watched me pass by.

What was going on in his head? He must have learned that it was me passing, that I was friendly and that I could make sounds just like he could. So what was it in his brain that made him think it was me? I mean…he had to have some picture in his brain's memory bank to compare me to other things that pass each morning…no? On my return jogs, the German shepherd looked at me from afar, as if waiting to see me again. He did not bark. Aristotle asserted that thoughts could not happen without images, so perhaps my canine friend was checking me out against his little brain's file of picture thoughts. How else could he think?

Wittgenstein tells us, "We make to ourselves pictures of facts."[1] For him, the picture is a model of what we take to be real. "The gramophone record, the musical thought, the score, the waves of sound, all stand to one another in that pictorial internal relation which holds between language and the world."[2]

When I think in words, those words are somehow being put into images, but maybe not. There is a difference between listening to radio and watching television. When I watch television, I see pictures. When I listen to radio, I make pictures.

When I do math, reading symbols, or just multiplying two numbers together in my head, I don't really picture the numbers, yet I think I do. What's going on?

It's as if I am thinking in cloudy symbols, although I don't really picture the cloudy symbols. The pictures are there, yet not there. When I try to picture a triangle, I do not have a clear picture of a triangle, just a vague fuzzy image, or more of an abstract representation of a triangle, some symbol that may not look exactly like a triangle, but something that stands in for the triangle. Why shouldn't that symbol be the triangle itself? What better symbol could there be?

We are all different. Some people are visual thinkers, others verbal, and still others may have thinking schemes beyond description. Nineteenth-century philologists and psychologists professed as "a matter of fact and not of argument" that thinking without verbal language was impossible.[3] We now know better. My "grrrrhoffing" friend may be thinking in smells.

The geneticist Francis Galton claimed that his thoughts almost never suggested words, and when those rare moments did suggest words, they were nonsense words like "the notes of a song might accompany thought. It often happens that after being hard at work, and having arrived at results that are perfectly clear and satisfactory to myself, when I try to express them in language I feel that I must begin by putting myself in quite another intellectual plane."[4]

As for words, the French mathematician Jacques Hadamard claimed, as did Galton, that words are neither followed by thoughts, nor thoughts by words:

> I insist that words are totally absent from my mind when I really think … even after reading or hearing a question, every word disappears at the very moment I am beginning to think it over; words do not reappear in my consciousness before I have accomplished or given up the research … and I fully agree with Schopenhauer when he writes, "Thoughts die the moment they are embodied by words."[5]

He goes on to say that this is also the case when he is thinking about algebraic symbols. He tells us that he represents general ideas as categorical circles A and B, such that if everything in A is in B then the circle A is imagined to be in B. If nothing in A is in B, then the two circles are distinct; there are no words involved, though it is possible that just the thought of any circle requires an instantaneous flash of the word.[6] Just imagine trying to logically make sense of one of Lewis Carroll's charming syllogisms that appeared in his symbolic logic book:

> No kitten that loves fish is unteachable.
> No kitten without a tail will play with a gorilla.
> Kittens with whiskers always love fish.
> No teachable kitten has green eyes.
> Kittens that have no whiskers have no tails.[7]

With a little work, we could construct a diagram of circles to untangle the syllogism. Without the circles or without some highly developed, nonverbal organizing scheme to illustrate to the mind the logic of these five sentences, how long would it take to conclude the logical consequence that "no green-eyed kitten will play with a gorilla"?

More revealing is Hadamard's presentation of his mental pictures of the steps in a proof that there are an unlimited number of prime numbers:

> In considering the primes from 2 to 11, he saw just "a confused mass."
> When forming the product of primes 2 to 11—that is, $2 \times 3 \times 5 \times 7 \times 11$—he imagined "a point rather remote from the confused mass."
> When he increased that product by 1, to get $(2 \times 3 \times 5 \times 7 \times 11) + 1$, he saw "a second point a little beyond the first."
> The number that he saw as a second point a little beyond the first, if not a prime, must contain a prime divisor, and therefore must be a prime larger than 11. This he saw as "a point somewhere between the confused mass and the first point."[8]

Michael Artin, my thesis advisor, would draw a squiggly figure looking roughly like

as if it were his initials, whenever he referred to what mathematicians call an "abelian variety." It had no graphical significance and certainly no resemblance to what an abelian variety really is. But to this day, I see that squiggly in my mind whenever I hear the words "abelian variety." It is, in many ways, Hadamard's confusing mass with transparent vagueness and useful hooks.

What is of interest here is that those representative points near and away from the confused mass have no properties of divisibility and no elements of prime-ness, just as Artin's squiggly drawing had no resemblance to an algebraic variety. As such, they play deceptive and fanciful roles in the conceptualizing process. So how does such a vague representational scheme help the logical process that fully depends on divisibility? Though Hadamard's images seemed to be devoid of divisor properties, it provided a mechanism for simultaneously viewing all the elements of the argument. He needed it to give the problem a face with trait and configuration, a character.

I have thought about this for some time, because Hadamard's answer is not satisfactory. The infinitude of prime numbers is one of the first real good proofs a young mathematician learns. A truly gifted young mathematician might prove it without looking at someone else's proof. The first proof I learned was that the square root of two is not rational; my second was that the prime numbers are infinite. So let me answer the question of what goes on in *my* mind when *I* rehash the proof that the prime numbers are infinite. Though my mathematical mind is nowhere near as sharp as Hadamard's was, my thoughts are remarkably similar to his in only one respect.

I too see the confused mass, as if it is just a jumble of things, as well as the representative points far and near. One difference is that I may be a bit more of a literal thinker and therefore have the actual symbols and peeking through the confused mass. You might say that my mass does not appear to be so confused. When it comes to the last step, looking for a prime divisor larger than 11, I see an actual number marked as a question mark somewhere between the confused mass and the first point. The reason, I suppose, is that—unlike a point—a question mark may mentally hold a division property, whereas a point (for me) is more of a placeholder.

The proof that $\sqrt{2}$ is irrational has a different mental flavor. It is what is called a proof by contradiction, and it goes like this: Assume $\sqrt{2}$ is rational, so $\sqrt{2} = \frac{p}{q}$, where p and q are integers and q is not equal to zero. Also assume that $\frac{p}{q}$ is in lowest terms, so that p and q cannot both be even numbers. Square both sides of that last equality so that $2 = \frac{p^2}{q^2}$. Multiply both sides by q^2 to show that $p^2 = 2q^2$ and therefore that p is an even number. Since p is even, it can be written as $2s$, for some integer s. Substitute $2s$ for p in the equality $\sqrt{2} = \frac{p}{q}$ so that $\sqrt{2} = \frac{2s}{q}$. Square both sides to get $2 = \frac{4s^2}{q^2}$. Then simplify to get $q^2 = 2s^2$, which means that q too must be even. This contradicts the assumption that p and q cannot both be even. QED.

Now here is what happens in my mind. I actually see the equation $\sqrt{2} = \frac{p}{q}$, as (I believe) anyone would. After all, what else would be the point of a symbolic equation? The squaring operation is also symbolic, but then I *see*, in my literal mind, a blob of many square numbers, starting with 4, 9, 16, 25, and 36, and ending with a blur of higher squares such as 100, 121, and 144. It's not as if the actual Indian numerals appear, but rather as if I just know the numbers are there. As soon as the equation is squared, the square root sign disappears and I suddenly see p in a circle of points that represents all the even numbers. The fact that p^2 is even doesn't come into play. My mind skips the step that links p^2 being even with p being even. Perhaps that is because I've used that connection so much in my life that I just don't bother to think about it; it has become part of my collected, undoubted knowledge.

Of course, the whole proof that $\sqrt{2}$ is irrational has also become part of my collected, undoubted knowledge—I have demonstrated it to freshmen thousands of times. So why isn't the proof itself imagined as a cloudy mass in the mind that could be made clear when necessary? It is. Like everything else well learned and repeated: some fringe-unconscious cloud may represent it as a sack containing everything that relates or associates to that simple proof by contradiction. It is a cauldron of forever-added perpetually simmering memories.

Who can know the inner thoughts or the inner struggles of a person's thinking schemes? It's hard enough to know our own. We may be able to know the components of the brain and their functions, what lights up red in an fMRI scan when a subject hears joyful news, or what lights up blue in a PET scan when a subject must make a risky decision. No matter how much we find out about the brain and how it works, we get no closer to really knowing how an individual thinks. Isn't that wonderful?

I used to ask my students how they thought of the alphabet, the seasons, or the months of the year. Every student had a different scheme. My alphabet is on a balance board centered at M. I suppose my last name has something to do with that. As I scan first the letters to the left of M, the board is tipped down on the left; once my scan passes M, the board tips downward on the right. It's as if I am on the board weighing down the letters with my scanning. Perhaps I learned how to spell my last name on a seesaw. But when I have to look up a word in the dictionary, or a name in the telephone book (paper dictionaries and telephone books, that is), there are no images. Strangely, I seem to know where I am and can zoom straight to the approximate place of the word or name before fine-tuning to get to the exact right place.

Months of the year are stranger still. On the ground is a great circle of tiles marked with the names of the months. I am standing on the tile that is exactly six months from the current month looking diametrically across the circle to that current month. It doesn't seem to matter whether I am standing on a tile six months ahead or six months behind. I have no idea where that bizarre calendar scheme comes from. None of my students have ever expressed their sense of calendar time in that way.

From what I just said, it would seem that I think in letters and words. Not so.

The words—if there at all—are a fog in the mind; the whole thinking process is so instantaneous that it is impossible to self-assess one's own thinking process. Words, pictures, or whatever else, may be part of the process, but they are like the individual notes of a piano concerto—once the bar is played, the ear has moved on.

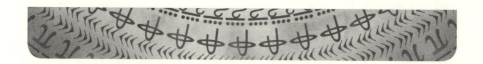

Chapter 24

Conclusion

> The words of language, as they are written or spoken, do not seem to play any role in my mechanism of thought. The physical entities which seem to serve as elements in thought are certain signs and more or less clear images.
>
> —Albert Einstein

We think in fuzzy pictures, cloudy symbols—there, yet not there—senses and impressions that allow us to go about our daily business. In literature, the conscious track has a lag time. Read Dostoyevsky's *Crime and Punishment* and come to the point when Raskolnikov crushes the old woman's head with a swing of an ax. What role does the ax play as we read further? Why did Dostoyevsky decide that the old woman should be killed by an ax and not by a gun nor bludgeoned to death with a poker? How would our psyches have responded if another weapon were used? The answer is in the skull bludgeoning. A skull smashing has connotations very different from a bruising to death. It leaves readers with contradictory emotions and clashing images in the mind: a gruesomely bloody death, and a humane swift death.

Analogously, in mathematics too, we have expressions that lead to competing perceptions, perhaps creating moods for the way we think. I have no way of showing this. So I can only offer my belief: that symbols have packaged teasers for unconscious suggestions playing in the background, while the mind rushes though hundreds of specific cases, instantly searching for connections to all the other times such an equation had been seen. Why not? The mathematics literature is filled with equations built from simple symbols. Those equations become symbols in their own right, offering powerful connections to the idea that something innocuous can occur over and over again in seemingly unrelated fields, sometimes relating the ephemeral to the physical.

Almost all thinking is multiple-tracked, with one track being conscious and another not. On one, there is plenty of action to keep the logic flowing. On another, there is the unconscious memory of everything that had been exposed to previous connections. And when more complicated equations are seen, a myriad of sympathetic thoughts come into play that are capable of conveying meaning through subconscious tracks that signal some connection to all those times the reader had seen a deeper fit with experience, or a deeper subconscious thought filled with creative possibilities.

Usually, reading is both a cognitive and an emotional activity. We read past words and phrases that may not consciously register as significant symbols, yet we find meaning at subliminal levels. We do not have to be aware of every word or phrase we read in order to grasp meaning. Meaning in literature comes from associative experiences. And so it is, at times, with reading mathematics and physics.

Unlike symbols in poetry, mathematical symbols begin as deliberate designs created by mathematicians. That does not stop symbols from performing the same function that a poem would: to make connections between experience and the unknown and to transfer metaphorical thoughts capable of conveying meaning.

As in poetry, there are archetypes in mathematics. If there are such things as self-evident truths, then there probably are things we know about the world that come with the human package at birth.

The 1950s baby experiments of Robert Frantz, Marc Bornstein, Eleanor Gibson, and other psychology researchers changed our impressions of how very young babies react to different patterns. They found that babies as young as eight weeks were already more interested in certain patterns than others. It was an exciting discovery at that time because it gave us evidence that from a very young age we begin to parse what we see in order to structure the world around us; infants are automatically captured by specific real-world structured patterns that stimulate their young nervous systems. In other words, the infant's interest behavior is directly controlled by structures in the world he or she will live in. Deep structures dictate our behavior and connect the dual tracks of our reading and thinking, the conscious and subconscious.

Plato's dialogue *Meno* refers to this. The dialogue is about whether or not virtue can be taught. But it uses the argument that the soul is immortal. To prove it, Socrates calls on a slave boy, an uneducated lad, for questioning. Through a series of questions and without help from anyone, Socrates manages to have the slave boy reason out facts about the Pythagorean theorem. Presumably, he was able to do this because he knew those truths before his birth and that they could be recalled from a previous life.

This is close to what Freud called "the collective subconscious of human phylogeny." Of course, Freud avoided the notion of soul. To him the soul is not spiritual, but rather the collective unconscious of the entire human species—that is, the unconscious memories passed from generation to generation through folklore, religion, and accumulated general knowledge about how to survive in a world of changing environment. This is how we come to know that *between two points there exists a unique line*, and that the presence of a serpent could mean death, the underworld, sex, fertility, sickness, or healing. The serpent doesn't appear in mathematics, except in two cases that I know of where serpentine diagrams indicate how to pick out entries on a grid. Yet, those self-evident truths that establish our axioms of arithmetic and geometry come from the collective subconscious of human phylogeny, the roots of inherited symbols.

Symbols transcend the medium of communication. They are ubiquitous in our language, and play a sizable role (though perhaps not a central one) in mathematical imagery linking the conscious and subconscious, the familiar and unknown, to give us cultural/emotional predispositions to meaning, all to enhance the creative process.

The entire mathematical phrase *between two points there exists a unique line* is no less a symbol than the first line of Frost's "The Road Not Taken." They are both distinguished by dominant subconscious powers coming from the collective subconscious. Though mathematical symbols are not generally found in the classical catalogue of archetypical symbols coming from folklore backgrounds, the serpent, the dove, the lion, and so on, they nevertheless encourage connections between the unknown and the familiar. In physics, we have Maxwell's equations: four interrelated equations that tell us how electric and magnetic fields relate to charge density and current density, and how they change with time. Like any great poem, Maxwell's equations tell us far more than what appears in the language. They form the basis for all electrodynamics and optics, and even lead to creative thinking about relativity and quantum mechanics.

In natural language, ordinary words describe what we see, think, or imagine. They have the power to create unfamiliar worlds and bring them into our imagination. Few special skills, other than those that come from being raised in a culture with other people, are required for admission to those worlds. To get to them, all we need is the experience of being human.

Mathematics is different. It usually requires a skill, sometimes a talent, often years of uncommon experience. I say usually, because there are many highly successful mathematicians and physicists who showed no apparent math aptitude at their young ages.[1] Though mathematics tends to use a symbolic language that bundles complexities of verbiage to simplify communication, it also draws on an astonishingly quick mental process that unpacks the essentials for making sense. And, like poetry, it uses a linguistic structure that enables readers to know the hidden meanings and the verbally unimaginable.

Typical symbols used in mathematics are operationals, groupings, relations, constants, variables, functions, matrices, vectors, and symbols used in set theory, logic, number theory, probability, and statistics. Individual symbols may not have much effect on a mathematician's creative thinking, but in groups they acquire powerful connections through similarity, association, identity, resemblance and repeated imagery. They may even create thoughts that are below awareness.

Whenever the form $\sqrt{x^2 + y^2}$ appears in an equation, the mathematician "knows" that it is representing a metric of some kind, possibly a distance in some coordinate system. It comes naturally from the Pythagorean theorem, which tells her that the distance from a point with coordinates (a, b) to a point with coordinates (c, d) is equal to $\sqrt{(c - a)^2 + (d - b)^2}$. In higher dimensions, the form could appear as $\sqrt{x^2 + y^2 + z^2}$, or as the square root of a sum of more than three squares. The form may not have come originally from any physical property, but the reader could interpret it through a geometric model that encourages associations. For example, the circle of radius R is given by the set of all coordinates (x, y) satisfying the equation $R = \sqrt{x^2 + y^2}$. It connects something ephemeral to a geometric image. Although $\sqrt{x^2 + y^2}$ is not generally found in the classical catalogue of archetype symbols with subconscious powers coming from folklore, it nevertheless encourages connections between the unknown and the familiar.

Some mathematical symbols begin as deliberate designs created to make connections between experience and the unknown, and to purposely transfer metaphorical thoughts capable of conveying meaning through analogy and resemblance. Others may accidentally do the same. For example, to represent the "sum" of a large but finite number of numerical terms, we use a large sigma Σ, the Greek "S" that has a finite number of discrete sharp corners (3, depending on how one counts). It likely comes from the Latin *summae* ("the sum of"). And when summing an infinite number of terms we use the S-shaped symbol $\int$ that is smooth and curved, suggestive of an infinite sum.

Meaning and understanding may be deeply embedded in associations and similitudes through experience and in the collective subconscious. Cultural predispositions of aesthetically persuasive symbols may play into our emotional appreciation of beauty, in mathematics as well as in poetry and art. Beauty in mathematics though—the elegance of proof, simplicity of exposition, ingenuities, simplification of complexities, making sensible connections—is, in a large part, attributable to the illuminating efficiency of smart and tidy symbols.

Appendix A

Leibniz's Notation

Ignoring minor technicalities, we may appreciate Leibniz's notation by taking the example $y = x^2$. We see that the numerical values of y depend on the values of x. If we pick some specific number—say, 2—then we may look at the ratio

$$\frac{x^2 - 2^2}{x - 2}.$$

The numerator factors neatly, so the ratio is really

$$\frac{(x-2)(x+2)}{(x-2)}.$$

At this point, we see that the denominator is the same as one of the factors of the numerator. Therefore, in the end we can say that the ratio we started with is really $x + 2$, as long as x is not equal to 2. But there is nothing special about the number 2. We could have done the same with any number—say, a. If we started with the ratio

$$\frac{x^2 - a^2}{x - a},$$

we would have ended with $x + a$, as long as x is not equal to a. Now, here comes the problem. We would like to know what happens to the ratio as $x - a$ approaches zero. We would like to know this because when $x - a$ approaches zero, the ratio tells us the rate at which x^2 is changing as x changes when we get very close to a. Of course, we cannot let $x - a$ become zero, because otherwise we could not perform the necessary cancellation to get $x + a$. The way around this is to look at what happens when $x - a$ approaches zero. That forces $x + a$ to approach $2a$. Since a was chosen to be an arbitrary number, it doesn't matter what a is; that means that the original ratio

$$\frac{x^2 - a^2}{x - a}$$

approaches a number that is really dependent just on a. It is what we call a function of a. The symbol $\frac{dy}{dx}$ represents that function, and it is called the "derivative of y with respect to x."

Why is $\frac{dy}{dx}$ such a good symbol? After all, the end result is not necessarily a ratio of two things. Our example ended up being $2a$, not a ratio at all.

For most problems of physical phenomena, you first know something about the rate of change of some function and then want to know the function itself—for example, you might know that $\frac{dy}{dx} = x$. Without questioning the unjustified symbolic manipulation, you would—as every calculus student is told to—think of $\frac{dy}{dx}$ as a fraction and multiply both sides by dx to get $dy = xdx$. How convenient. It turns out that those strange little variables dx and dy actually do follow the rules of algebra synthetically: if y is a function of x and in turn x is a function of t, then $\frac{dy}{dx}\frac{dx}{dt} = \frac{dy}{dt}$. And if both x and y are functions of t, then

$$\frac{\frac{dy}{dt}}{\frac{dx}{dt}} = \frac{dy}{dx}.$$

This is a setup for Leibniz's other brilliant symbol, the "integral." The integral operates on a function. For the sake of simplicity, once again we use an example—say, $y = x$. The integral operating on y gives us the function that has y as a rate of change.

It turns out that if two functions are equal, then their integrals differ by just one constant number. The integral symbol in this case is $\int ydx$. So, if we take the integral of both sides of the equation $dy = xdx$, we symbolically get $\int dy = \int xdx$. The left side is asking for that function that has 1 as a rate of change with respect to the variable y. That must be just y itself. The right side is asking for that function that has x as a rate of change with respect to the variable x. That turns out to be $\frac{x^2}{2}$. Therefore, $y - \frac{x^2}{2} = C$, where C is some numerical constant.

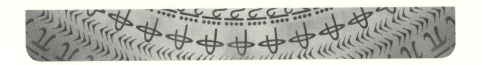

Appendix B

Newton's Fluxion of x^n

Let the Quantity x flow uniformly, and let the Fluxion of x^n be to be found. In the same time that the Quantity x by flowing becomes $x + o$, the Quantity x^n will become $\overline{x + o}\,^n$—that is, by the Method of Infinite Series

$$x^n + nox^{n-1} + \frac{nn - n}{2} oox^{n-2} + \&\text{c.}$$

And the Augments

$$o \text{ and } nox^{n-1} + \frac{nn - n}{2} oox^{n-2} + \&\text{c.}$$

Are to one another as

$$1 \text{ and } nx^{n-1} + \frac{nn - n}{2} ox^{n-2} + \&\text{c.}$$

Now let those Augments vanish and their ultimate Ratio will be the Ratio of 1 to nx^{n-1}; and therefore the Fluxion of the Quantity x is to the Fluxion of the Quantity x^n as 1 to nx^{n-1}.

Taken from John Harris's 1723 English translation of Isaac Newton's *Introductio ad De Quadratura Curvarum*.[1]

Appendix C

Experiment

Here is a transcript of an interview designed as an experiment in symbolic cognition performed at the Joint Meetings of the American Mathematical Society and the Mathematical Association of America in Boston on January 4, 2012.

At the center of my laptop screen was the following figure:

$$\sqrt{\left(\frac{y}{2}\right)^2 + z^2 + \frac{y}{2}}.$$

The typical interview would go something like this:

Q: What goes through your mind when you see such a thing as this (pointing to the screen of my laptop)?

A: Well…(Long pause).

Q: There's no right or wrong answer. I just want to know how you are seeing this.

A: There's a sum of squares under the radical [the square root sign], so this might have something to do with an ellipse. … No, … wait. It's a cone that has a hyperbolic cross section in one direction and a parabola in another.

At this point, the equation $x^2 + bx + c = 0$ would fade in at the top of the screen with two arrows pointing downward to the center figure. For a full 10 seconds, the screen displayed:

$$x^2 + bx + c = 0$$

$$\downarrow$$

$$\sqrt{\left(\frac{y}{2}\right)^2 + z^2} + \frac{y}{2}.$$

The subjects were looking directly at the screen, when, after 10 seconds, the equation and arrows faded out.

I interviewed nine people this way; all but two tried to tie the question to the graph of the equation in question. But after the strange 10-second display of the fading-in-and-out equation, two of my interviewees got the same idea. The following is in effect the transcript of one interview. The other is virtually identical.

A: Hold on, maybe the x, y, and z are not variables.

There was no verbal indication that they were aware of the fade-in/fade-out equation on the page, but in each of the two interviews, the interviewee wrote

$$\sqrt{\left(\frac{b}{2}\right)^2 + c^2} + \frac{b}{2}$$

on a pad. It was as if the second interviewee had seen what the first had done.

Q: Mm-hmm. So what are you seeing now?
A: Looks something like…a solution to the general quadratic equation? Is it the positive root of a quadratic equation?
Q: What do you think?

The subject wrote $x^2 + bx + c = 0$.

A: No-no, … (rewriting the equation replacing $+b$ with $-b$).

Rewriting the equation replacing the $+c$ with $-c$, the subject continued to look at the new equation, while I said nothing. Finally after a few moments, he wrote $x^2 - bx - c^2 = 0$. With a surety that made him continue, he finally wrote $x^2 - yx - z^2 = 0$.

Q: Nice!

A: (Eyes widening.) "There, ... this (pointing to the figure at the center of the laptop screen) is the positive value of x in the equation $x^2 - yx - z = 0$.

I designed the problem to be harder than I should have by using a z^2 rather than a z. My purpose was to make the terms under the square root sign take the form of an ellipse, just to complicate things. I could have just started with

$$\sqrt{\left(\frac{b}{2}\right)^2 + c^2} + \frac{b}{2}$$

instead of

$$\sqrt{\left(\frac{y}{2}\right)^2 + z^2} + \frac{y}{2},$$

but I thought that that would be a give away.

In the end, I asked each subject if he or she had seen anything unusual on the screen of my laptop while they were contemplating the question. Everyone, including the exceptional two, claimed to have seen nothing fade in or out.

Two people saw the

$$\left(\frac{b}{2}\right)^2 + \text{something}$$

under the radical, which reminded them of the form $b^2 - 4c$ that always appears under a radical when trying to find the solution to an equation of the form $x^2 + bx + c = 0$. Graphically, the form

$$\sqrt{\left(\frac{b}{2}\right)^2 + c^2}$$

could also suggest a positive elliptical cone. What made two subjects see

$$\sqrt{\left(\frac{y}{2}\right)^2 + z^2} + \frac{y}{2}$$

as the solution to a quadratic equation and seven subjects search for a graphical connection, when all nine must have subconsciously imbibed the hinted connection with $x^2 + bx + c = 0$?

I cannot tell you in words how I process knowing that

$$\sqrt{\left(\frac{y}{2}\right)^2 + z^2 + \frac{y}{2}}$$

is the positive root of the quadratic equation $x^2 + bx + c = 0$. I just see it through the process of seeing that anything of the form

$$\left(\frac{y}{2}\right)^2 + 4 \times \text{something}$$

under a square root symbol reminds me of the form $b^2 - 4c$ and hence of the root of $x^2 + bx + c = 0$.

Appendix D

Visualizing Complex Numbers

Mark the complex number $a + ib$ as if it were a point in the Cartesian plane as (a, b). In that way, all complex numbers that happen to be real numbers lie along the horizontal line through $(0, 0)$, and all complex numbers that happen to be imaginary numbers lie along the vertical line through $(0, 0)$ (figure D.1).

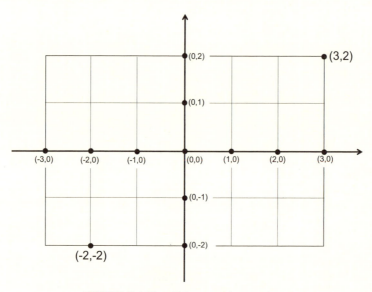

FIGURE D.1 Visualizing complex numbers.

Every complex number is represented as a pair of numbers and pictured in this plane. But why are we calling these numbers, when they seem to be pairs of numbers? The answer is that they obey the rules of arithmetic for numbers. Add any two, and you get a third: define $(a, b) + (c, d)$ as $(a+c, b+d)$. [Notice that $(a+c, b+d)$ is the representative point of $(a + ib) + (c + id)$, which is $(a+c) + i(b+d)$.] What about multiplication? We define the product of (a, b) with (c, d) to be $(ac - bd, ad + bc)$. [Notice that $(ac - bd, ad + bc)$ is the representative point of $(a + ib)(c + id)$, which is $(ac - bd) + i(ad + bc)$.] With these definitions of addition and multiplication, all the laws of arithmetic are satisfied without contradiction. But something interesting happens when we look further. Multiplication has a meaning. Multiplication by i is a 90-degree counterclockwise rotation. Multiplication by di is a 90-degree counterclockwise rotation followed (or preceded) by a scaling factor of d.

All of this could have been said using notation that kept $\sqrt{-1}$ instead of the new representative i, which has the same virtual meaning. But i isolates the concept of rotation from the perception of root extraction, offering the mind a distinction between an algebraic result and an extension of the idea of number.

Appendix E

Quaternions

What William Rowand Hamilton realized was that he could write $x + iy + jz + kw$ as if each term is independent of the others with the multiplication rule $i^2 = j^2 = k^2 = ijk = -1$ and that it would be represented as quadruples (x, y, z, w) with a multiplication rule that obeyed all the laws of algebra except the commutative rule. He would have to be content with the fact that $ij = k$, and $ji = -k$. He would have to accept more than two square roots of -1, infinitely more! He would have to accept that quadratic equations have more than two solutions. These are some of the trade-offs for extending numbers to a higher dimension beyond the complex. The new system includes the complex numbers. Embedded is a three-dimensional imaginary system represented by quaternions of the form $iy + jz + kw$.

So what does multiplication by any of the i, j, k's do in the three-dimensional space? Rotate, we expect. How? If i, j, k denote three positive unit directions of the mutually perpendicular axes in three-dimensional space, then multiplying j by i rotates the entire three-dimensional space by 90 degrees, sending the i-axis to the j-axis while holding the k-axis fixed. It tells us that three-dimensional space has two distinct orientation models and that the physicist must decide which should be conventional. In other words, should the grooves of woodscrews be designed to enter wood by clockwise turning, or by counterclockwise turning? The choice is arbitrary, but convention favors clockwise. If you studied physics in college, you may remember these rotations as the right-hand rule for the orientation of space, an understanding that is elemental to both physics and mathematics.

Unlike the complex numbers, the quaternions have no representation in a space that we know and can visualize. They are not open to comfortable visualization by those of us who are untrained to see in four dimensions. Yet we include them as legitimate numbers in our generalized number system. They didn't come to us by way of geometry. They may have come as a result of symbolic representation. Now they turn up in the most unexpected places. Had Euler not marked $\sqrt{-1}$ as i in a memoir presented to the Academy at St. Petersburg in 1777, had $\sqrt{-1}$ not been published as i in 1794 after his death, had Gauss not made consistent use of i after 1801, the quaternions might not have been discovered so soon in the history of the subject to make their vital contributions to mathematical physics.

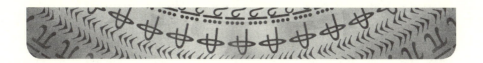

Acknowledgments

This book could not have been written without the support of many individuals, starting with my wife, Jennifer, who read several complete drafts, making editorial, structural, and practical suggestions. She is always my inspiration and support.

I am exceedingly grateful to the Bogliasco Foundation for a Fellowship and Residency to complete this book at the magnificent Villa dei Pini on the aptly named *Golfo Paradiso* in Bogliasco, Italy. It generously provided me with ample time, luxurious comfort, and late afternoon swims in the warm Mediterranean, all in the company of my delightfully inspiring colleagues Ruby Blondell, John Eaton, Luisa Costa Gomes, Jennie MaryTai Liu, Jennifer Sachs, and Willard Spiegelman, who contributed both directly and indirectly to the final draft of this book.

Very special thanks to the caring readers of complete drafts: Michelle Bower, Julian Ferholz, and Margorie Senechal. Thanks to others who read drafts of individual chapters, often many chapters: Robyn Aariarhod, Steve Batterson, Kenneth Bleeth, Charles Burnett, Barry Cipra, David Cox, Robert Dawson, Florin Diacu, Rafaella Franci, Fernando Gouvea, Emily Grosholz, Phil Holmes, Jens Høyrup, Gizem Karaali, Mikhail Katz, Catherine Mazur-Jefferies, Barry Mazur, Peter Meredith, Kim Plofker, and Siobhan Roberts. And thanks to my expert consultants for informative and encouraging correspondence, and conversations: Stanislas Dehaene, David Geary, Daniel Kahneman, George Lakoff, Steven Pinker, Ian Stewart, David Tall, and Elisabetta Ulivi.

I consider myself a mathematics journalist embedded with the learned troupes. So I thank the real scholars who have labored over the history of symbols to uncover what had been intellectually lost for hundreds of years. Stephen Chrisomalis for his doctoral dissertation, "The Comparative History of Numerical Notation" (McGill, 2003), which lists 1,047 academic publications about numerals and numeral systems, numeral notation, and numeration. This was vastly helpful in my research. Paolo Vian of the Biblioteca Apostolica Vaticana. Clay Institute, Project Guttenberg, OpenLibrary.org, University of Chicago Digital Preservation Collection, The European Cultural Heritage Online (ECHO), The New York Public Library Digital Gallery, Liberty Fund, PhilSci Archive, Biblioteca Apostolica Vaticana, Archivio di Stato de Firenze, Biblioteca Marucelliana Firenze, Biblioteca Nazionale Centrale Firenze, Universita degli Studi di Pavia, and so on. Gallica (the online rare books library of the Bibliothèque National de France), Scribd.com, Ancientlibrary.com, the Perseus Digital Library, Centro di Ricerca Matematica Ennio De Giorgi (for Bombelli's *L'Algebra*), and Biblioteca della Scuola Normale Superiore for allowing me to do research from my home that, a decade ago, would have taken years in rare books rooms of libraries halfway around the world, and to Jonathan Bennett, who maintains a site for the translations of correspondences of mathematicians and philosophers at www.earlymoderntexts.com. And to the real scholars no longer with us, who labored over the history of symbols to uncover what had been intellectually lost for hundreds of years: Florian Cajori, Sir Thomas Heath, Louis Charles Karpinski, and Karl Menninger. Many thanks to Marek Čtrnáct, the translator for the Czech edition, for catching a heap of English typos in the last moments of production.

Special thanks goes to my editor, Vickie Kearn, for her resolute support, and to my agent, Andrew Stewart, who saw this project's potential in my very brief proposal.

To the Mazur Jefferies family: Catherine, Tom, Sophie, Yelena, and Ned. To the Marshall family: Tamina, Scott, and Lena. To my brother Barry and sisters-in-law Carole Joffe and Grechen Mazur, for their constant encouragement.

Notes

Introduction

1. Eugene Wigner, "The Unreasonable Effectiveness of Mathematics in the Natural Sciences, Richard Courant Lecture in Mathematical Sciences Delivered at New York University, May 11, 1959," *Communications in Pure and Applied Mathematics*, vol. 13, no. I (1960): 1–14.

2. My thanks to Robert Dawson for pointing out that Wigner's story is a retelling of a much older story told in de Morgan's *A Budget of Paradoxes* (New York: Cosmo Classics, 2007), 285–286.

3. There is the biblical reference to the priest's bathing pool in Solomon's Temple (Kings 7:23), "And he made a molten sea, ten cubits from the one brim to the other: it was round all about, and his height was five cubits: and a line of thirty cubits did compass it round about." It leads to the interpretation that $\pi = 3$. But this is not a true reference to π as any sort of constant.

4. The verb "to syncopate" as used in this book means to shorten a word by omitting letters from its middle. It is a specific form of abbreviation, although most abbreviations are not syncopations. In the middle of the nineteenth century, the German G.H.F. Nesselman characterized the development of algebra notation by three stages, which he called rhetorical, syncopated, and symbolic—in that order. For more about this, see John Wesley Young, William Wells Denton, and Ulysses Grant Mitchell, *Lectures on Fundamental Concepts of Algebra and Geometry* (Norwood, MA: Norwood Press, 1911), 226.

5. Mohammed Ben Musa al-Khwārizmī, *The Algebra*, trans. and ed. Frederic Rosen (London: Oriental Translation Fund, 1831), 41–42.

6. I thank Fernando Gouvea for pointing this out to me.

Definitions

1. *Webster's Third New International Dictionary of the English Language Unabridged.*

2. Ibid.

Chapter 1: Curious Beginnings

1. Piotr Wojtal and Krzyszgof Sobxzyk, "Man and Woolly Mammoth at the Krakow Spadzista Street (B)—Taphonomy of the Site," *Journal of Archaeological Science*, vol. 32, no. 2 (2005): 193–206.

2. Quentin Atkinson, R. D. Gray, and A. J. Drummond, "mtDNA Variation Predicts Population Size in Humans and Reveals a Major Southern Asian Chapter in Human Prehistory," *Molecular Biology and Evolution*, vol. 25, no. 2 (2008): 468–474.

3. Joseph Campbell, *The Hero with a Thousand Faces* (Novato, CA: New World Library, 2008), 22.

4. Joseph Campbell, *The Power of Myth* (New York: Anchor, 1991), 4–5.

5. A.W.G. Pike, D. L. Hoffmann, M. García-Diez, P. B. Pettitt, J. Alcolea, R. De Balbín, C. González-Sainz, C. de las Heras, J. A. Lasheras, R. Montes, and J. Zilhão, "U-Series Dating of Paleolithic Art in 11 Caves in Spain," *Science*, vol. 336, no. 6087 (2012): 1409–1413.

6. Marc D. Hauser, Noam Chomsky, and W. Tecumseh Fitch, "The Faculty of Language: What Is It, Who Has It, and How Did It Evolve?" *Science*, vol. 22, no. 298 (5598) (2002): 1569–1579. Michael Tomasello, *Origin of Human Communication* (Cambridge, MA: MIT Press 2008). There is a long-standing debate on whether or not animals have language symbols that is both imperative and declarative to signal a want for something such as food, hugs, or something be-

yond, such as conversational exchange of information for its own sake. I do not take a side in this debate. For more on this debate, see Sue Savage-Rumbaugh, Stuart Shanker, and Talbot Taylor, *Apes, Language and the Human Mind* (Oxford, UK: Oxford University Press, 1998); and "Missing Links in the Evolution of Language," in *Characterizing Conciousness: From Cognition to the Clinic?*, ed. Stanislas Dehaene and Yves Christen (Berlin: Springer, 2011), 1–26.

7. David A. Anthony, *The Horse, the Wheel, and Language: How Bronze Age Riders from the Eurasioan Steppes Shaped the Modern World* (Princeton, NJ: Princeton, 2007), 67.

8. James P. Allen, *Middle Egyptian: An Introduction to the Language and Culture of Hieroglyphics* (Cambridge, UK: Cambridge University Press, 2010), 2.

9. Ibid., 3.

10. Joseph Campbell, *The Power of Myth*, xviii.

11. H. G. Wells, *The Outline of History*, vol. 1 (Garden City, NY: Garden City Books, 1956), 172.

12. Edwin C. Krupp, *Echoes of the Ancient Skies: The Astronomy of Lost Civilizations*, Astronomy Series (New York: Dover, 2003), 62–70.

13. D. E. Smith, *History of Mathematics*, vol. 1 (New York: Dover, 1958), 7.

14. Levi Leonard Conant, *The Number Concept: Its Origins and Development* (New York: Macmillan, 1931), 5. Also see Daniel G. Brinton, *Essays of an Americanist* (Philadephia: David McKay, 1890), 406.

15. According to James Henry Breasted, the earliest dated event in history (4241 BC) is the Egyptian calendar. It is a twelve-month, thirty-days-per-month calendar that is ostensibly better than our current one. There is some doubt that this Egyptian calendar is really that old. It may have dated only as far back as the First Egyptian Dynasty, King Djer (ca. 3000 BC). See James Henry Breasted, *Ancient Times* (Boston: Ginn & Co., 1966).

16. Joseph Campbell, *Myths to Live By* (New York: Penguin, 1993), 8.

Chapter 2: Certain Ancient Number Systems

1. Daniel Zohary and Maria Hopf, *Domestication of Plants in the Old World: The Origin and Spread of Cultivated Plants in West Asia, Europe, and the Nile Valley* (Oxford, UK: Oxford University Press, 2001), 241–243.

2. Bill T. Arnold, *Who Were the Babylonians?* (Leiden: Brill Academic Publishing, 2005), 2.

3. Eleanor Robson, "Mesopotamian Mathematics," in *The Mathematics of Egypt, Mesopotamia, China, India and Islam*, ed. Victor Katz (Princeton, NJ: Princeton University Press, 2007): 64.

4. It stayed in Plimpton's private collection until his death in 1936. It was bequeathed to Columbia University, where it remains in the Cuneiform Collection.

5. E. J. Banks, "Description of Four Tablets Sent to Plimpton," unpublished manuscript, Columbia University Libraries Special Collection, Cuneiform Collection, no date.

6. Eleanor Robson, "Neither Sherlock Holmes nor Babylon: A Reassessment of Plimpton 322," *Historia Mathematica* 28 (2001): 167–206.

7. Early cuneiform markings were made with reeds, leaving circular or semicircular impressions.

8. In a way, just two symbols 𓍢 and 𓏤 were used. The other 58 symbols could be considered as concatenations of the two basic symbols; however, the mind would see each distinct concatenation differently.

9. Florian Cajori, *A History of Mathematical Notations* (New York: Dover, 1993), 12.

10. Ibid., 23–25.

11. This number is often confused with Aristarchus's later estimate 10^{63} using Archimedes's method.

12. Otto Neugebaur, *The Exact Sciences in Antiquity* (New York: Dover Publications, 1969), 10–11.

13. Sterling Dow, "Greek Numerals," *American Journal of Archaelogy*, vol. 56, no. 1 (1952): 21–23.

14. Cajori, *A History of Mathematical Notations*, 32.

Chapter 3: Silk and Royal Roads

1. Joseph Dauben, "Chinese Mathematics," in *The Mathematics of Egypt, Mesopotamia, China, India, and Islam: A Sourcebook*, ed. Victor Katz (Princeton, NJ: Princeton University Press, 2007), 297.

2. Ibid. 189.

3. Lay Yong Lam, "The Development of Hindu-Arabic and Traditional Chinese Arithmetic," *Chinese Science*, vol. 13 (1996): 35–54.

4. For an abridged English translation, see Lay Yong Lam, "Nine Chapters on the Mathematical Art: An Overview," *Archive for History of Exact Sciences*, vol. 47, no. 1 (1994): 2–51.

5. Dirk J. Struik, *A Concise History of Mathematics* (New York: Dover Publications, 1987), 74.

6. Wann-Sheng Horng, "Euclid versus Liu Hui: A Pedagogical Reflection," in *Using History to Teach Mathmatics: An International Perspective*, ed. Victor Katz (Cambridge, UK: Cambridge University Press, 2000), 44.

7. Translation by Joseph Dauben found in Victor Katz, ed., *The Mathematics of Egypt, Mesopotamia, China, India, and Islam*, 230.

8. Lay Yong Lam and Tian Se Ang, *Fleeting Footsteps: Tracing the Conception of Arithmetic and Algebra in Ancient China* (Hackensack, NJ: World Scientific, 2004), 182.

9. See the English translation in Lay Yong Lam and Tian Se Ang, *Fleeting Footsteps*, 149–182.

10. Robert Temple, *The Genius of China: 3,000 Years of Science, Discovery, and Invention* (New York: Simon & Schuster, 1986), 141.

11. Lay Yong Lam and Tian Se Ang, *Fleeting Footsteps*, 49.

12. Suan Shu Shu, trans. Joseph Dauben, ed. J. C. Marzloff, "The Suan Shu Shu (A Book on Numbers and Computation), A Preliminary Investigation," *Archive for History of Exact Sciences*, vol. 62, no. 2 (2008): 91–178.

13. Yan Li and Shi Ran Du, *Chinese Mathematics: A Concise History*, trans. John N. Crossley and Anthony W.-C. Lun (Oxford, UK: Clarendon Press, 1987). Yoshio Mikami, *The Development of Mathematics in China and Japan* (New York: Chelsea, 1974). Philip D. Straffin, "Liu Hui and the First Golden Age of Chinese Mathematics," *Mathematics Magazine*, vol. 71, no. 3 (1998): 163–181.

14. Lay Yong Lam and Tian Se Ang, *Fleeting Footsteps*, 183.

15. For well-described, explicit directions on how the other operations of arithmetic were carried out, see Joseph Dauben's essay "Chinese Mathematics."

16. Lay Yong Lam and Tian Se Ang, *Fleeting Footsteps*, 185.

17. D. E. Smith and J. Ginsburg, "From Numbers to Numerals and from Numerals to Computation," in *The World of Mathematics*, vol. I, ed. J. R. Newman (New York: Simon and Schuster, 1956), 442–463.

18. From *Epinomis*, trans. A. E. Taylor, ed. Edith Hamilton and Huntington Cairns (Princeton, NJ: Princeton University Press, 1961), 978b, 1521.

19. Ibid., 977c, 1520 978b.

Chapter 4: The Indian Gift

1. G. R. Kay, "Notes on Indian Mathematics," *Journal and Proceedings of the Asiatic Society of Bengal*, vol III (1907): 475–508.

2. Kim Plofker, "Mathematics in India," in *The Mathematics of Egypt, Mesopotamia, China, India, and Islam: A Sourcebook*, ed. Victor Katz (Princeton, NJ: Princeton University Press, 2007), 386.

3. Lay Yong Lam, "Linkages: Exploring the Similarities between the Chinese Rod Numeral System and Our Numeral System," *Archive for History of Exact Sciences*, vol. 37, no. 4 (1987): 365–392.

4. G. G. Joseph, *The Crest of the Peacock, Non-European Roots of Mathematics* (Princeton, NJ: Princeton University Press, 2000), 215–216.

5. Tobias Dantzig, *Number: The Language of Science*, ed. Joseph Mazur (New York: Plume, 2007), 19.

6. "Gobar" is the name given to the West Arabic numerals. The word "Gobar"

originates from the Arabic *ghubar*, meaning "dust." It refers to the ancient form of reckoning in dust or sand.

7. It is worth noting that many of the authorities cited in these last few paragraphs drew their information from vastly different sources, few of which would be considered reliable by present-day historians.

8. Henry Burchard Fine, *The Number-System of Algebra: Treated Theoretically and Historically* (Boston: D. C. Heath, 1903), 90.

9. Kay, "Notes on Indian Mathematics," 475–508.

10. For the full finger counting scheme, see James Gow, *A Short History of Greek Mathematics* (Cambridge, UK: Cambridge University Press, 1884), 25.

11. J. Wassmann and P. R. Dasen, "Yupno Number System and Counting," *Journal of Cross-Cultural Psychology*, vol. 25, no. 1 (1994): 78–94.

12. D. E. Smith, *History of Mathematics* (New York: Dover, 1958), 197.

13. For further details and an interesting excursion into how finger counting was used by traders in different cultures, see Karl Menninger, *Number Words and Number Symbols: A Cultural History of Numbers* (New York: Dover, 1992), 201–220.

14. This works because $ab = (a - 5 + b - 5) \times 10 + (10 - a) \times (10 - b)$.

15. The reason for this comes from the algebraic identity $ab = ((a-10)+(b-10)) \times 10 + (a - 10) \times (b - 10) + 100$. A similar method works for any two numbers but gets more complicated because more than five fingers may have to be raised on each hand. The idea is to make use of the formula $ab = ((a-c)+(b-c)) \times c + (a - c) \times (b - c) + c^2$, where c is the amount the numbers should be reduced in order to bring the multiplication down to a manageable number. Unfortunately, larger numbers require knowing how to square c.

16. D. E. Heath, *History of Mathematics* (New York: Dover, 1953), 119–120.

17. Brian Butterworth, *What Counts: How Every Brain Is Hardwired for Math* (New York: Free Press, 1999).

18. W. Penfield and T. Rasmussen, *The Cerebral Cortex of Man* (New York: Macmillan, 1952).

19. There are only two known Roman abacuses. One is at the Cabinet des Médailles (Bibliothèque nationale de France) in Paris; the other is at the Museo delle Terme Diocleziano in Rome.

20. The second groove from the right has five counters to indicate multiples of an ounce, each representing one ounce. The counter in the shorter groove represents six ounces. The first groove on the right is divided into three parts, to indicate a half-ounce, a quarter-ounce, and a third of an ounce.

21. See Charles Burnett, "The Abacus at Echternach in ca. 1000 AD," *Sources and Commentaries in Exact Sciences*, vol. 3 (2002): 91–108.

22. *Apices* is a term whose meaning is confusing when seen in High Middle Age texts on calculation. It sometimes means the Gerbertian abacus itself, also known as the *arcs of Pythagoras* (even though Pythagoras surely did not contribute to such an idea). At other times, it more appropriately meant the counters themselves, which were were conical and hence had apexes. And still at other times, it meant the symbols that appeared on the counters. For the remainder of this book, it will mean the physical counters themselves.

23. Menninger, *Number Words and Number Symbols*, 324.

24. Burnett, "The Abacus at Echternach in ca. 1000 AD," 92.

Chapter 5: Arrival in Europe

1. Augustus De Morgan, *Elements of Algebra* (London: Taylor and Walton, 1837), i.

2. Sigler, Laurence E., trans., *Fibonacci's Liber Abbaci* (New York: Springer-Verlag, 2002).

3. Albrecht Heeffer, "The Abbaco Tradition (1300–1500): Its Role in the Development of European Algebra," Working paper, 2008, pp. 1–2.

4. Thanks to Van Egmond's extensive cataloguing work in the 1970s, we now have a list of about 250 abbaco manuscripts. The list appears in Warren Van Egmond, "Practical Mathematics in the Italian Renaissance: A Catalogue of Italian Abacus Manuscripts and Printed Books to 1600," *Monografia*, 4 (Florence: Istituto e Museo di Storia della Scienza, 1980).

5. J. B. Mullinger, *The Schools of Charles the Great and the Restoration of Education in the Ninth Century* (Chicago: Norwood 1980), 12.

6. Before the fifteenth century, the word "abacus" referred to the broader meaning of calculation beyond the mechanical tool. It referred to "doing arithmetic."

7. Jens Høyrup, *Jacopo da Firenze's Tractatus algorismi and Early Italian Abacus Culture* (Basel: Birkhauser, 2007), 44.

8. This translation, taken from the Latin preface of Fibonacci's *Liber abbaci*, appears on page 8 in Professor Albrecht Heeffer's paper at the Centre for Logic and Philosophy of Science, Ghent University, titled "Epistomic Justification and Operational Symbolism," available at http://logica.ugent.be/albrecht/thesis/ EpistemicJustification.pdf (accessed August 13, 2013).

9. Jens Høyrup, "Leonardo Fibonacci and Abbaco Culture: A Proposal to Invert the Roles," *Revue d'histoire des mathématiques*, vol. 11 (2005): 23–56.

10. D. E. Smith and L. C. Karpinski, *The Hindu-Arabic Numerals* (Boston: Ginn and Co., 1911), iii.

11. Thomas Frank, Steven John Livesey, and Faith Wallis, *Medieval Science, Technology, and Medicine* (London: Routledge, 2005), 135.

12. Warren van Egmond, "The Commercial Revolution and the Beginnings of Western Mathematics in Renaissance Florence, 1300–1500," Ph.D. Thesis, Indiana University, 1976.

13. D. E. Smith and Yekuthiel Ginsburg, "Rabbi Ben Exra and the Hindu-Arabic Problem," *American Mathematical Monthly*, vol. 25 (1918): 99–108.

14. W. W. Rouse Ball, *A Short Account of the History of Mathematics* (London: Macmillan, 1908), 168.

15. Al-Biruni, *Alberuni's India: An Account of the Religion, Philosophy, Literature, Geography, Chronology, Astronomy, Customs, Laws and Astrology of India about A.S. 1030*, ed. Edward C. Sachau, vol. II (London: Trubner & Co, 1888), 15.

16. Jan P. Hogendijk and Abdelhamid I. Sabra, eds., *The Enterprise of Science in Islam: New Perspectives (Dibner Institute Studies in the History of Science and Technology)* (Cambridge, MA: MIT Press, 2003), 3–18.

17. There are several good books on the history of zero, so there is no need to go into any detail here. I recommend Robert Kaplan, *The Nothing That Is: A Natural History of Zero* (New York: Oxford University Press, 2000); and Charles Seife, *Zero: The Biography of a Dangerous Idea* (New York: Penguin, 2000).

18. Radha Charan Gupta, "India," in *Writing the History of Mathematics: Its Historical Development*, ed. Joseph Warren Dauben and Christopher J. Scriba (Basel: Birkhäuser, 2002), 307.

19. W. W. Rouse Ball, *A Short Account of the History of Mathematics*, 144–146.

Chapter 6: The Arab Gift

1. The *al* in an Arabic name means "from the birthplace of." Khwarazm is a province in present-day Uzbekistan.

2. Carl B. Boyer and Uta C. Merzbach, *A History of Mathematics* (New York: John Wiley, 2011), 228.

3. By the tenth century, there were numerous Arabic texts on the Indian numerals. For example:

   •Abu l'Hasan Al Uqlidisi, *Kitab al fusal fil hisab al Hindi* (950 AD)

   •Abu'l Hasan al-Qifti, *Chronology of the Scholars* (900 AD), which includes a part on Indian place-value systems

   •Al-Kindi, *Ketab fi Isti'mal al-'Adad al-Hindi* (*On the Use of the Indian Numerals*) (ca. 830 AD).

   •Abul Wafa Al-Buzjani, *Kitab al-Hindusa* (*Science of the Indians*) (940–997, 998 AD)

   •Kushyer Ibn Labban, *Kitab Fi Usual Hisab Al Hind* (971 AD); this is the earliest extant Arabic book on Indian numerals

   For further details, see Georges Ifrah, *The Universal History of Numbers* (New York: Wiley, 2000), 589. Also see Abu Kamil, *Principles of Hindu Reckoning*, trans. Martin Levey (Madison: University of Wisconsin, 1966), 24.

4. Robert of Chester's Latin translation of the *Algebra of al-Khowarizmi* (New York: Macmillan, 1915).

5. For a comprehensive list of early translations of al-Khwārizmī's *Algebra*, see Albrecht Heeffer, "A Conceptual Analysis of Early Arabic Algebra," in *Unity of Science in the Arabic Tradition*, ed. Shahid Rahman, Tony Street, and Hassan Tahiri (New York: Springer-Verlag, 2008), 91–92.

6. A thirteenth-century Latin translation is at the Cambridge University Library under MS Ii, vi.5, 104r–111v.

7. Fibonacci, *Liber abbaci*, trans. L. E. Sigler (New York: Springer-Verlag, 2002), 17.

8. Charles Burnett, "Learning Indian Arithmetic in the Early Thirteenth Century," *Boletín de la Asociación Matemática Venezolana*, vol. IX, no. 1 (2002): 15.

9. Otto Neugebaur, *The Exact Sciences in Antiquity* (New York: Dover Publications, 1969), 24n.

Chapter 7: *Liber Abbaci*

1. See pages 435, 430, 273, and 274 in Fibonacci, *Liber abbaci*.

2. Ibid., 1.

3. For more detail on this subject, see Charles Burnett, *Numerals and Arithmetic in the Middle Ages* (Farnum, Surrey, UK: Ashgate, 2010), xi, 87–97; or Charles Burnett, "Fibonacci's 'Method of the Indians,'" *Bollettino di Storia delle scienze matematiche*, vol. 23 (2003 [published 2005]): 87–97.

4. Burnett, "Learning Indian Arithmetic in the Early Thirteenth Century," 91.

5. I am very grateful to Raffaella Franci for pointing this out to me.

6. Raffaella Franci, "Trends in Fourteenth-Century Italian Algebra," *Oriens-Occidens*, vol. 4 (2004): 81–105.

7. Elisabetta Ulivi, "Benedetto da Firenze (1429–1479), un maestro d'abbaco del xc secolo. Con documenti inediti e con un'Appendice su abacisti e scuole d'abaco a Firenze nei secoli xiii–xvi," *Bollettino di Storia delle Scienze Matematiche*, no. 22 (2002): 1–243.

8. I do not know precisely when this book was discovered, but surely sometime before 1989. Some popular books have attributed the discovery to Raffaella Franci. However, in my conversations with Franci, she claimed that the book was actually discovered by Arrighi.

9. Gino Arrighi, "Maestro Umbro (Sec. XIII) Livero De L'Abbecho (cod. 2404 della Biblioteca Riccardiana di Firenze)," *Bollettino Della Deputazione Di Storia Patria Per L'Umbria*, vol. LXXXVI (1989): 5–140.

10. Jens Høyrup, "Leonardo Fibonacci and Abbaco Culture: A Proposal to Invert the Roles," *Revue d'histoire des mathèmatiques*, vol. 11 (2005): 27–28.

11. Two other treatises written in Pisa in the early thirteenth century, now being studied by Franci, might give clues to the contents and treatment of Fibonacci's lost *Liber minoris guise*. See Raffaella Franci, "Leonardo Pisano e la trattatistica dell'abaco in Italia nei secoli XIV e XV," *Bollettino di Storia delle scienze matematiche*, vol. 23, no. 2 (2003): 33–54.

12. Ibid., 82.

13. I thank Raffaella Franci for clarifying the debate on the influence of Fibonacci on the spread of Indian numerals in Italy.

14. From correspondence with Charles Burnett.

15. Høyrup, "Leonardo Fibonacci and Abbaco Culture," 23.

16. A large number of manuscript copies of the Carmen de Algorismus are still around, so we know it must have been popular. See James Andrew Corcoran, Patrick John Ryan, and Edmond Francis Prendergast, "The Catholic Church and the Gentle Science of Numbers," *American Catholic Quarterly Review*, vol. 44 (1919): 135.

17. In the twelfth century, the library of the Salem Abbey was one of the most important in Europe.

18. Smith and Karpinski, *The Hindu-Arabic Numerals*, iii. Georges Ifrah, *The Universal History of Numbers* (New York: Wiley, 2000), 556–566. A convenient annotated translation of the major part of Sacrobosco's text can be found in *A*

Source Book in Medieval Science, ed. E. Grant (Cambridge, MA: Harvard University Press 1974), 94–102.

19. A. L. Basham, *The Wonder That Was India: A Survey of the Culture of the Indian Sub-Continent before the Coming of the Muslims* (New Delhi: Picador, India edition, 2005), 414.

20. Thomas F. Glick, Steven Livesey, and Faith Wallis, eds., *Medieval Science, Technology, and Medicine: An Encyclopedia* (Routledge Encyclopedias of the Middle Ages) (Oxford, UK: Routledge, 2005), 39.

21. Charles Burnett, "The Semantics of Indian Numerals in Arabic, Greek, and Latin," *Journal of Indian Philosophy*, vol. 34 (2006): 15–30.

22. Richard Lemay, "The Hispanic Origin of Our Present Numeral Forms," in *Viator*, vol. 8: *Medieval and Renaissance Studies*, ed. Henry Ansgar Kelly (Los Angeles: University of California Press, 1977), 435–438.

23. El-Mas'údí, *Meadows of Gold*, vol. 1., trans. Aloys Sprenger (London: Oriental Translation Fund, 1841), 201.

24. Ibid., 27.

25. Clemênt Huart, *A History of Arabic Literature* (London: William Heinemann, 1903), 183.

26. El-Mas'údí, *Meadows of Gold*, 200–201.

27. I'm told that the oldest surviving Indian numerals book, *Livero del abbecho*, dates back to around 1290, but I have not seen this book. It is listed in Warren Van Egmond, *Practical Mathematics in the Italian Renaissance: A Catalogue of Italian Abbacus Manuscripts and Printed Books to 1600* (Florence: Instituto e Museo di Storia Della Scienza, 1981).

28. George Peacock, "History of Arithmetic," in *Encyclopedia Metropolitana*, ed. Samuel Taylor Coleridge, London: (1847), 369–523.

    C. A. Bayly, *Empire and Information: Intelligence Gathering and Social Communication in India, 1780–1870* (Cambridge, UK: Cambridge University Press, 1996). Also in Kapil Raj, "Colonial Encounters and the Forging of New Knowledge and National Identities: Great Britain and India, 1760–1850," *Osiris*, vol. 15 (Nature and Empire: Science and the Colonial Enterprise, 2000): 119–134. And Kapil Raj, *Relocating Modern Science: Circulation and the Construction of Scientific Knowledge in South Asia and Europe, 17th and 19th Centuries* (Delhi: Permanent Black, 2006).

Of course, other reasons should be taken in account as well, such as the fact that Peacock could have only been in literate-in-English circles.

See the essay Charles Burnett, "Indian Numerals in the Mediterranean Basin in the Twelfth Century, with Special Reference to the Eastern Forms," in *From China to Paris: 2000 Years' Transmission of Mathematical Ideas (Boethius. Texte und Abhandlungen zur Geschichte der Mathematik und der Naturwissenschaften)*, ed. Benno van Dalen, Joseph Dauben, Yvonne Dold-Semplonius, and Menso Folkerts (Wiesbaden: Franz Steiner Verlag, 2002), 240–245.

Chapter 8: Refuting Origins

1. Michael Farquhar, *A Treasury of Deception* (New York: Penguin, 2005), 150–151.
2. Ken Alder, "History's Greatest Forger: Science, Fiction, and Fraud along the Seine," *Critical Inquiry*, vol. 30 (Summer 2004): 704–716.
3. Their arguments are partially documented in BNF Res-Z-1249/livres rares, at the Bibliothèque nationale in France. (I have not seen the document.)
4. Guillaume Libri, *Histoire des science en Italie: depuis la renaissance des lettres jusqu'à la fin du dix- septième siècle*, vol. 1 (Paris: Jules Renouard, 1838), 117–135.
5. Mémoires et Communications, *Comptes Rendus Hebdomadaires des Séances de l'Académie des Sciences*, vol. 12 (1841): 741–756.
6. Agathe Keller, at the University of Paris VII-CNRS, found 99 titles on the origin of numbers and arithmetic published between 1827 and 1907, and believes the list is not exhaustive.
7. G. R. Kaye, "Notes on Indian Mathematics—Arithmetical Notations," *Journal of the Asiatic Society of Bengal*, n.s. vol. III , no. 7 (1907): 475–508.

8. The birch bark leaves are now in the Bodleian Library at Oxford (MS. Sansk. d. 14), but are currently too fragile to be examined.

9. G. R. Kaye, "Notes on Indian Mathematics," 493.

10. Bibhutibhusan Datta, "Review: G. R. Kaye, The Bakhshali Manuscript—A Study in Medieval Mathematics," *Bulletin of the American Mathematical Society*, vol. 35, no. 4 (1929): 579–580. See also Bibhutibhusan Datta, "The Bakhhshali Manuscript," *Bulletin of the Calcutta Mathematical Society*, vol. 21 (1929): 1–60.

11. G. G. Joseph, *The Crest of the Peacock, Non-European Roots of Mathematics* (Princeton, NJ: Princeton University Press, 2000), 215–216.

12. This is from a paper appearing in the archives of the Centre Pour la Communication Scientifique, which does not appear to be a peer reviewed journal. Available at http://hal.archives-ouvertes.fr/docs/00/45/53/92/PDF/PeacockAK.pdf (accessed August 13, 2013).

13. Benoy Kumar Sarkar, *Hindu Achievements in the Exact Sciences* (Ithaca, NY: Cornell University Library [scanned from the 1918 edition], 2009), 8–11.

14. G. R. Kaye, "Notes on Indian Mathematics," 293–297.

15. Bibhutibhusan Datta, "The Bakhshali Manuscript," 1–60.

16. Karl Menninger, *Number Words and Number Symbols: A Cultural History of Numbers* (New York: Dover, 1992), 406.

17. Charles Burnett, "Learning Indian Arithmetic in the Early Thirteenth Century," *Boletín de la Asociación Matemática Venezolana*, vol. IX, no. 1 (2002): 15–26.

18. Benno van Dalen, Joseph Dauben, Yvonne Dold-Samplonius, and Menso Folkets, eds., *China to Paris: 2000 Years' Transmission of Mathematical Ideas (Boethius. Texte und Abhandlungen zur Geschichte der Mathematik und der Naturwissenschaften)* (Wiesbaden: Franz Steiner Verlag, 2002), 266.

19. Menninger, *Number Words and Number Symbols*, 315.

20. Orstein Ore, *Number Theory and Its History* (New York: Dover, 1988), 21.

21. See G. F. Hill, *The Development of Arabic Numerals in Europe* (Oxford, UK: Oxford University Press, 1915), 28.

Chapter 9: Sans Symbols

1. See http://www.claymath.org/library/historical/euclid (accessed August 13, 2013).

2. Euclid, II, 7.

3. Proclus, *A Commentary on the First Book of Euclid's Elements*, trans. with introduction and notes by Glenn R. Morrow (Princeton, NJ: Princeton University Press, 1992).

4. Stacy Schiff, *Cleopatra: A Life* (New York: Little Brown, 2010), 67–68.

5. Alberto Manguel, *A History of Reading* (New York: Penguin, 1996), 43.

6. Sir Thomas Heath, *Diophantus of Alexandria: A Study in the History of Greek Algebra* (Cambridge, UK: Cambridge University Press, 1910), 32–34.

7. Ibid., 41–42.

8. Frederic Rosen, *The Algebra of Mohammed Ben Musa*, ed. and trans. Frederic Rosen (London: Printed for the Oriental Translation Fund, 1831), 10–11. Also see accessible version found in Florian Cajori, *A History of Mathematical Notations* (New York: Dover, 1928), 84.

9. Jens Høyrup, "Hesitating Progress—The Slow Development toward Algebraic Symbolization in Abbacus and Related Manuscripts, ca. 1300 to ca. 1550," in *Philosophical Aspects of Symbolic Reasoning in Early Modern Mathematics, Studies in Logic*, vol. 2, no. 26, ed. Albrecht Heeffer and Maarten Van Dyck (London: College Publications, 2010), 3–56.

10. By the quadratic formula, the solution is calculated as

$$x = \frac{10}{2} \pm \sqrt{\left(\frac{10}{2}\right)^2 - 21}$$
$$= 5 \pm \sqrt{25 - 21}$$
$$= 5 \pm \sqrt{4}$$
$$= 5 \pm 2$$
$$= 3 \ or \ 7.$$

11. Høyrup, "Hesitating Progress," 3–56.

12. Ibid.

13. Tobias Dantzig, *Number: The Language of Science*, ed. Joseph Mazur (New York: Plume, 2007), 90. Jacob Klein argued that something was lost in the liberation. See Jacob Klein, *Greek Mathematical Thought and the Origin of Algebra* (New York: Dover, 1992).

Chapter 10: Diophantus's *Arithmetica*

1. Ian Stewart, *Why Beauty Is Truth: A History of Symmetry* (New York: Basic Books, 2007), 34.

2. We see this in *Arithmetica*, book V, problem 2: to find three numbers in geometical progression such that each when added to a given number gives a square. He picks the given number to be 20 and is forced to solve $4x^2 + 20 = 4$. He then says that it is absurd (ατοπον), because the 4 had better be some number greater than 20.

3. Sir Thomas L. Heath, *A History of Greek Mathematics*, vol. II (Oxford, UK: Clarendon, 1921), 458.

4. L. D. Reynolds and N. G. Wilson, *Scribes and Scholars: A Guide to the Transmission of Greek and Latin Literature* (Oxford, UK: Clarendon Press, 1978), 50.

5. D'Arcy Thompson, "The S of Diophantus," *Transactions of the Royal Society of Edinburgh*, vol. XXXVIII, no. 17 (1896): 607–609. Also see James Gow, *History of Greek Mathematics* (Cambridge, UK: Cambridge University Press, 1884), addenda, ix.

6. Sir Thomas L. Heath, *Diophantus of Alexandria: A Study in the History of Greek Algebra* (Cambridge, UK: Cambridge University Press, 1910), 62–64.

7. Note that these abbreviations are also the first syllables.

8. D'Arcy Thompson, "The S of Diophantus," 607–609.

9. James Gow, *History of Greek Mathematics*, addenda, 109 footnote.

10. E. A. Wallis Budge, *Egyptian Hieroglyphic Dictionary* (Whitefish, MT: Kessinger Publishing, 2003).

11. Heath, *Diophantus of Alexandria*, 32–36.

12. Available at http://books.google.com/books?id=RL1CAAAAcAAJ&printsec=frontcover&source=gbs_ge_summary_r&cad=0#v=onepage&q&f=false (accessed August 12, 2013).

13. James Gow, *History of Greek Mathematics*, addenda, 109 footnote.

14. The oldest (Vat. gr. 191) dates from the thirteenth century. The others, from the fourteenth, fifteenth, and sixteenth centuries, are also in the Biblioteca Apostolica Vaticana under Barb. gr. 267, Pal. gr. 391, Reg. gr. 128, Ross. 980, Urb. gr. 74, Vat. gr. 200, Vat. gr. 304.

15. The translation can be found at Qusta ibn Lukqa, trans., *The Arabic Text of Books IV to VII of Diophantus's Arithmetika* (Ann Arbor, MI: University Microfilms, 1979).

16. Now, thanks to Google e-Books, Bachet's entire edition of Diophantus's *Arithmetica* may be viewed online at http://books.google.com/ebooks/reader?id=RL1CAAAAcAAJ&printsec=frontcover&output=reader (accessed August 12, 2013).

17. John W. Baldwin, *The Scholastic Culture of the Middle Ages, 1000–1300* (Lexington, MA: D. C. Heath, 1971), 40.

18. I suggest the translation of *ars rei et census* to be "the art and quality of 'the thing'"; I suspect that by *rei*, "the thing," he must have meant "the unknown."

19. Heath, *Diophantus of Alexandria*, 21.

20. Ibid., 20.

21. Ibid., 20.

22. Tannery, Paul, *Dictionary of Scientific Biography* (New York: Scribner, 1970), 251–257. Also see Sir Thomas L. Heath, *Diophantus of Alexandria*, 15.

23. Thanks to European Cultural Heritage Online, the entire manuscript of Wilhelm Xylander's translation of book 6 may be found at http://echo.mpiwg-berlin.mpg.de/ECHOdocuViewfull?start=1&viewMode=images&ws=1.5&mode=imagepath&url=/mpiwg/online/permanent/library/W770Y3H9/pageimg&pn=1 (accessed August 12, 2013).

24. Marcianus 308 is appropriately at the Biblioteca Marciana in Venice.

25. In the Bachet translation, the characters are lowercase and the x-square is this strange symbol: 𝒷𝒶, but in other translations the symbols are in capitals.

26. This was the notation that was used in Latin translations of many ancient manuscripts from the mid-fifteenth century to the mid-seventeenth century. See W. W. Rouse Ball, *A Short Account of the History of Mathematics* (London: Macmillan, 1908), 216.

27. Bachet, *Arithmetica*, 321.

28. Diophanti Alexandrini, *Opera Omnia*, vol. 1, ed. and trans. into Latin by Paulus Tannery (Leipzig: B.G. Teubneri, 1843), xxxiv–xxxix.

29. Diophanti Alexandrini, *Opera Omnia*, 6–7.

30. Bachet VI, problem 12. Also see Heath, *A History of Greek Mathematics*, 460; and Gow, *History of Greek Mathematics*, 110, for simple fractions.

31. This is James Gow's translation in Gow, *History of Greek Mathematics*, addenda, 108. Heath has a more accurate translation in Heath, *Diophantus of Alexandria*, 129: "Perhaps the subject will appear rather difficult, inasmuch as it is not yet familiar (beginners are, as a rule, too ready to despair of success); but you, with the impulse of your enthusiasm and the benefit of my teaching, will find it easy to master; for eagerness to learn, when seconded by instruction, ensures rapid progress."

32. We also appreciate the breakup of large numbers in groups of three by commas. Such a scheme is found in Fibonacci's *Liber abbaci*, although, rather than commas, he uses things akin to paretheses to group the digits.

Notice that in the Bachet translation, the symbols are all lowercase. Bachet talks of 1Q.+2N.+1 as the translation of .

Chapter 11: The Great Art

1. *Encyclopaedia of the History of Science, Technology, and Medicine in Non-Western Cultures*, ed. Helaine Selin (Dordrecht: Kluwer Academic, 1997), 162.

2. Mohammed Ben Musa al-Khwārizmī, *The Algebra of Mohammed Ben Musa*, ed. and trans. Frederic Rosen (London: J. Murray, 1831), 5.

3. (1) The fourteenth-century Vienna MS. (Codex Vindobonensis 4770 Rec. 3246 XIV. 339.80). (2) The fifteenth-century Dresden MS. (Codex Dresdensis C. 80). (3) The sixteenth-century Columbia University MS. (Codex Universitatis Columbiae, MS X 512, Sch. 2, Q.)

4. Charles Hutton, *Mathematical and Philosophical Dictionary*, vol. 1 (London: J. Johnson, 1796), 63.

5. Louis Charles Karpinski, *Robert of Chester's Latin Translation of the Algebra of Al-Khwarizmi* (New York: Macmillan, 1915), 111. This book also has both Latin and English translations.

6. Ibid., 111.

7. Ibid., 9.

Chapter 12: Symbol Infancy

1. Samuel Johnson, *Dictionary of the English Language, and An English Grammar*, 6th ed. (London: Revington, Payne, etc., 1785), 318.

2. Michael Sean Mahoney, *The Mathematical Career of Pierre de Fermat, 1601–1665* (Princeton, NJ: Princeton University Press, 1994), 32.

3. E. T. Bell, *The Development of Mathematics* (New York: Dover, 1945), 34.

4. Ibid., 125.

5. A. Djebbar, *Enseignement et recherche mathématiques dans le Maghreb des XIIIe –XIVe siècles* (Paris: Université Paris-Sud, Publications Mathématiques d'Orsay, 1981), 55–75.

6. For details on how the substitution $y = x - \frac{a}{3}$ reduces the general cubic equation to one with no quadratic term, see Paul J. Nahin, *An Imaginary Tale: The Story of* $\sqrt{-1}$ (Princeton, NJ: Princeton University Press, 1999), 8–11.

7. For more of the story on the famous feud, see John Derbyshire, *Unknown Quantity: A Real and Imaginary History of Algebra* (New York: Joseph Henry, 2006), 81–85.

8. Girolamo Cardano, *Ars Magna or the Rules of Algebra*, trans. T. Richard Witmer (New York: Dover, 1993), 8.

9. Ibid., 96.

10. Albrecht Heeffer, "Negative Numbers as an Epistemic Difficult Concept: Some Lessons from History," Working paper, Center for Logic and Philosophy of Science, Ghent University, 5. No year given.

11. Ibid., 6.

12. Cardano, *Ars Magna or the Rules of Algebra*, 9.

13. Ibid.

14. Ernst Mach, *Popular Scientific Lectures*, trans. Thomas J. McCormack (Chicago: Open Court, 1895), 195–196.

Chapter 13: The Timid Symbol

1. Codex Gotting. Philos. 30, University of Göttingen. The author refers to himself as Initius Alegbras.

2. W. W. Rouse Ball, *A Short Account of the History of Mathematics*, London: Macmillan, 1908), 215. Ball claims that the cube root was marked as ✔✔✔ and the fourth root as ✔✔, but I could not find any such notation in *Die Coss*, other than the notation ✔.✔ to indicate something very different indeed. He also claims the year of publication to be 1526; however, the only published edition I could find—the one edited by Michael Stifel—is dated 1554. Michael Stifel, *Die Coss Christoffe Ludolffs mit schönen Exempeln der Coss. Gedrückt durch Alexandrum Lutomyslensem* (Königsberg, Prussia [now Kaliningrad, Russia], 1554).

3. Stifel, *Die Coss Christoffe Ludolffs*, folio 83.

4. Ibid., folio 125.

5. This could also have come from the Latin word *radix* ("root").

6. Cajori, *A History of Mathematical Notations*, vol. I, 366–369.

7. Jens Høyrup, "Hesitating Progress—The Slow Development toward Algebraic Symbolization in Abacus and Related Manuscripts, ca. 1300 to ca. 1550," Conference paper of *Philosophical Aspects of Symbolic Reasoning in Early Modern Science and Mathematics*, Ghent, 27–29 (August 2009): 18.

8. Ibid., 19.

9. Nicolas Chuquet, *Renaissance Mathematician: A Study with Extensive Translation of Chuquet's Mathematical Manuscript Completed in 1484*, ed. Graham Flegg, Cynthia Hay, Barbara Moss (Dordtrecht: Reidel, 1985), 93.

10. Ibid.

11. The copy of Summa that I saw is a translation into old German with old spelling and words packed so closely together that it is hardly readable. So I relied on Florian Cajori's *A History of Mathematical Notations*, vol. I (New York: Dover, 1993), 336.

12. G Beaujouan, "The Place of Nicolas Chuquet in a Typology of Fifteenth Century French Arithmetic," in *Mathematics from Manuscript to Print 1300–1600*, ed. C. Hay (Oxford, UK: Oxford University Press, 1988), 73–88. Also see B. Moss, "Chuquet's Mathematical Executor: Could Estienne de la Roche Have Changed the History of Algebra?" in ibid., 117–126.

13. Cajori, *A History of Mathematical Notations*, vol. I, 100.

14. The idea of negative exponents can be found in John Wallis, *Mathesis Universalis* (Oxford, UK: Oxford, 1657), 65–68. Isaac Newton perfected the idea of negative exponents. Negative exponents occur in his "De quadratura Cervarum," and can be found in *The Mathematical Papers of Isaac Newton*, vol. VII, 1691–1695, ed. D. T. Whiteside (Cambridge, UK: Cambridge University Press, 1976), 154.

15. W. W. Rouse Ball, *A Short Account of the History of Mathematics* (London: Macmillan, 1908), 217.

Chapter 14: Hierarchies of Dignity

1. From the preface of Bombelli's *L'Algebra* found in Barry Mazur and Federica La Nave, "Reading Bombelli," *The Mathematical Intelligencer*, vol. 24, no. 1 (2002): 12–21.

2. Florian Cajoli claimed that he had communication from the Bolognian mathematician Ettore Bortolotti that the equal sign was developed at Bologna "independent of Robert Recorde and perhaps earlier." See Florian Cajori, *A History of Mathematical Notations*, vol. 1, 126.

3. I searched through the entire *L'Algebra* to find hundreds of cases where the words *fa, faro, eguali*, and *eguale* are used, but few instances where two expressions were bound by the word *equale*, as it would have been used in saying anything like "1+1 è eguale a 2."

4. Rafael Bombelli, *L'Algebra*, book II (Bologna: Giouanni Rossi, 1579), 204. Thanks to Biblioteca della Scuola Normale Superiore and Centro di Ricerca Matematica Ennio De Giorgi, this can be found at http://mathematica.sns.it/opere/9/ (accessed August 13, 2013).

5. William Shakespeare, *Cymbeline, King of Britain*, act IV, scene 2.

6. Barry Mazur, *Imagining Numbers: Particularly the Square Root of Minus Fifteen* (New York: Farrar, Straus and Giroux, 2003), 107–131.

7. Florian Cajori, *A History of Mathematical Notations*, vol. II, 127.

8. Ibid., 224

Chapter 15: Vowels and Consonants

1. François Viète, *Opera mathematica* (Leiden: Elzevir, 1646), 399.

2. Note that the symbol π denoting the ratio of the circumference to the diameter of a circle was an idea of William Jones, who used it for the first time in 1706. To find out more about the process, see Petr Beckmann, *A History of Pi* (New York: St. Martins, 1976), 92–94.

3. In 1646, an assemblage of Viète's works was published and edited by Francisci

van Schooten, the editor of Viète's *Opera*, who included commentaries and notes using the symbol $\sqrt{}$ with a vinculum (overbar) that can be extended over several terms. It is interesting to note that van Schooten got the formula for $2/\pi$ wrong when trying to translate Viète's rhetorical description into symbolic notation on page 400. It shows how easy it is to make mistakes when trying to interpret what words in a phrase mean.

4. Occasionally Viète uses Recorde's symbol for the equal sign. See Viète, *Opera mathematica*, 20.

5. Viète, *Opera mathematica*, 246. Available at http://books.google.com/ebooks/reader?id=DAGlicM_CkMC&printsec=frontcover&output=reader (accessed August, 13, 2013).

6. Michael Sean Mahoney, *The Mathematical Career of Pierre de Fermat, 1601–1665* (Princeton, NJ: Princeton University Press, 1994), 32.

7. Euclid, *Elements*, vol. I, trans. Sir Thomas Heath (New York: Dover, 1956), 373.

8. Ibid., p. 376.

9. $(a + b)^2 = (a + b)(a + b) = a(a + b) + b(a + b) = aa + ab + ba + bb = a^2 + 2ab + b^2 = a^2 + b^2 + 2ab.$

10. Tobias Dantzig, *Number: The Language of Science*, ed. Joseph Mazur (New York: Plume, 2007), 90.

11. I thank Fernando Gouvea for pointing this out to me.

12. Stevin argued in favor of accepting irrationals as numbers in his *L'Arithmetique* (1585).

13. Florian Cajori, *A History of Mathematical Notations*, vol. II (New York: Dover, 1993), 131.

14. It actually says a bit more than this—namely, that any polynomial of degree $n \geq 1$ with complex coefficients has at least one solution. But if it has one solution, then it is easy to see that it must have n solutions. This comes from knowing that the polynomial splits into a product of two polynomials, one of them being $(x - r)$, where r is the root that was guaranteed by the fundamental theorem, and the other a polynomial of degree $n - 1$.

15. Viète, *Opera mathematica*, 12.

Chapter 16: The Explosion

1. For a deeper understanding of Descartes's *Geometry*, see Emily R. Grosholz, *Representations and Productive Ambiguity in Mathematics and the Sciences* (Oxford, UK: Oxford University Press, 2007), 165–183.

2. René Descartes, *Geometria a Ranato Descartes*, ed. Florimondi de Beaune and Francisci van Schooten (Amsterdam: Elzevir 1659), 2.

3. Many thanks to Emily Grosholz for pointing this out.

4. René Descartes, *Geometria*, ed. F. van Schooten (Leiden: Elzevir, 1659), 3.

5. Florian Cajori, *A History of Mathematical Notations*, vol. I (New York: Dover, 1993), 374.

6. Descartes, *Geometria*, 4.

7. Ibid., 69.

8. Ibid., 6.

9. I thank Emily Grosholz for pointing out to me the seventeenth-century difficulty with the idea that nature is mechanical and mathematical.

10. A. P. Youschkevitch, "The Concept of Function up to the Middle of the 19th Century," *Archive for History of Exact Sciences*, vol. 16, no. 1 (1976): 37–85.

Chapter 17: A Catalogue of Symbols

1. John Aubrey, *"Brief Lives," Chiefly of Contemporaries, Set Down by John Aubrey, between the Years 1669 & 1696*, vol. II, ed. Andrew Clark (Oxford, UK: Clarendon Press, 1898), 108.

2. Florian Cajori, *A History of Mathematical Notations*, vol. I, 250.

3. Ibid.

4. A. N. Whitehead, *An Introduction to Mathematics* (Oxford, UK: Oxford University Press, 1958), 60.

5. Ibid., 59.

6. Ibid., 200.

7. Ibid., 244.

8. Ball, *A Short Account of the History of Mathematics*, 241.

9. Ibid., 243. (In looking through the *Ars Conjectandi*, I could not find a single instance where Bernoulli uses the symbol ∞.)

10. This originally appeared in William Oughtred's *Clavis mathematicae* (1648), and can be more easily found in Cajori, *A History of Mathematical Notations*, vol. I, 427.

11. Ibid., vol. II, 130–131.

Chapter 18: The Symbol Master

1. A quote from John Theodore Merz, a Leibniz biographer: Hal Hellman, *Great Feuds in Science, Ten of the Liveliest Disputes Ever* (New York: John Wiley & Sons, 1998), 41.

2. Translation from *Die Philosophischen Schriften*, ed. C. I. Gerhardt, vol. VII, Berlin 1875–90, 22, found in Margaret E. Baron, *The Origins of the Infinitesimal Calculus* (Oxford, UK: Pergamon, 1969), 9.

3. Typesetters also objected to fractions being written as $\frac{a}{b}$. They preferred $a : b$, which was also used, but fortunately the three-terraced type won out.

4. Cajori, *A History of Mathematical Notations*, vol. I, 182–183.

5. Alessandro Padoa, *La Logique Deductive* (Paris: Gauthier-Villars, 1912), 21.

Chapter 19: The Last of the Magicians

1. Hellman, *Great Feuds in Science*, 41.

2. John of Salisbury, *The Metalogicon: A Twelfth-Century Defense of the Berbal and Logical Arts of the Trivium*, trans. Daniel McGarry (Baltimore: Paul Dry Books, 2009), 167.

3. James R. Newman, ed., *The World of Mathematics*, vol. II (New York: Simon and Schuster, 1956), 140.

4. Tobias Dantzig, *Number: The Language of Science*, ed. Joseph Mazur (New York: Plume, 2007), 135.

5. Fritjof Capra, *The Turning Point* (London: Fontana Collins, 1983), 49.

6. Isaac Newton, *The Mathematical Papers of Isaac Newton*, vol. VII, 1691–1695, ed. D. T. Whiteside (Cambridge, UK: Cambridge University Press, 1976).

7. Isaac Newton, *The Mathematical Papers of Isaac Newton*, vol. VIII, 1697–1722, ed. D. T. Whiteside (Cambridge, UK: Cambridge University Press, 2008), 123.

8. Mikail Katz and David Sherry, "Leibniz's Infinitesimals: Their Fictionality, Their Modern Implementations, and Their Foes from Berkeley to Russell and Beyond," *Erkenntnis* 78, no. 13 (2013): 572–625.

9. I learned this from Mikhail Katz, who read this chapter with care and corrected several of my misunderstandings of Leibniz and Newton.

10. Of course, we realize that as long as o was considered to be chosen as a very small quantity, the end result could not be exact, but only an approximation of the exact. Both Newton and Leibniz were not thinking in our more modern concept of "limits," though they were certainly close.

11. Bishop George Berkeley, *The Analyst; Or, A Discourse Addressed to an Infidel Mathematician* (originally printed in London for J. Tonson in 1734); William B. Ewald, ed., *From Kant to Hilbert: A Source Book in the Foundations of Mathematics* (Oxford, UK: Oxford University Press, 1996), 60–92. Available at http://www.maths.tcd.ie/pub/HistMath/People/Berkeley/Analyst/Analyst.html (accessed August 13, 2013).

12. Some mathematicians have argued that Berkeley was wrong. See Katz and Sherry, "Leibniz's Infinitesimals."

13. Joseph Mazur, *The Motion Paradox* (New York: Dutton, 2007), 151–152.

Chapter 20: Rendezvous in the Mind

1. A. N. Whitehead, *An Introduction to Mathematics* (Oxford, UK: Oxford University Press, 1958), 40–41.

2. Sir William Molesworth, ed., *The English Works of Thomas Hobbes of Malmesbury*, vol. VII, lesson V (London: Longman, Brown, Green and Longmans, 1845), 329–330.

3. It was in a rebuttal essay addressed as: To the same egregious professors of the mathematics, one of geometry, the other of astronomy, in the chairs set up by the noble and learned Sir Henry Savile, in the University of Oxford. It was a hostile essay that even ridiculed the professors' Latin grammar. The intended professors were John Wallis and Seth Ward—both held Salvillian Chairs at that time.

4. I thank David Tall for pointing this out. See David Tall and Scholmo Vinner, "Concept Image and Concept Definition in Mathematics with Particular Reference to Limits and Continuity," *Educational Studies in Mathematics*, vol. 12, no. 2 (1981): 151–169. The concept of definition is the topic in David Tall's book. See David Tall, *How Humans Learn to Think Mathematically: Journeys through Three Worlds of Mathematics* (Cambridge, UK: Cambridge University Press, 2013).

5. Darcy Wentworth Thompson, *On Growth and Form*, ed. John Tyler (Cambridge, UK: Cambridge University Press, 1992), 269.

6. I owe this point to Robyn Arianrhod, who expressed it in her wonderful book, *Einstein's Heroes* (Oxford, UK: Oxford University Press, 2005), 133.

7. Historically, that curious thing $\sqrt{-1}$ came about by questioning cubic equations, not quadratic equations. For a more inclusive story of that history, see Paul Nahin, *An Imaginary Tale: The Story of $\sqrt{-1}$* (Princeton, NJ: Princeton University Press, 1998), 8–11.

8. We may also think of it as the geometric mean between +1 and −1. In other words, +1 is to i as i is to −1. Symbolically, that would be represented as $+1 : i = i : -1$.

9. Whitehead, *An Introduction to Mathematics*, 64.

10. Ian Stewart, *Why Beauty Is Truth: A History of Symmetry* (New York: Basic Books, 2007), 152.

Chapter 21: The Good Symbol

1. William Oughtred, *Clavis Mathematicae* (Strasbourg: Johann Crosleu and Amos Curteyne, 1657), 66. Note: Florian Cajori tells us that in 1652, William Oughtred represented the ratio of the circumference to the diameter of a circle as $\frac{\pi}{\delta}$, not separating the meaning of π or δ, though, presumably the π stood for *periphe-ria* (periphery), and δ stood for *diametrum* (diameter). My reading of the *Clavis* suggests otherwise: on page 66, there is a clear separation of π and δ (see the following figure, from Google Books).

2. Cajori, *A History of Mathematical Notations*, vol. 2, p. 9.

3. Benjamin Peirce was in favor of something quite different. He knew of the close connection between π and e, the base for the natural logarithm, and felt that that connection should be part of the symbolism. He proposed ⨅ for π and ⨆ for e, true symbols. How fortunate that his proposal didn't take; just imagine what a burden those symbols would be for persons with dyslexia. The remarkable Euler identity would look like ⨆⨅^{i} +1 = 0. Does it mean $e^{i\pi} +1 = 0$ or $\pi^{ie} +1 = 0$?

4. In fact, $i^i = e^{-\frac{\pi}{2}} \approx 0.207879576$. The way to see this is to write

$$i = \cos\frac{\pi}{2} + i\sin\frac{\pi}{2} = e^{i\frac{\pi}{2}}.$$

   Raise both left and right sides by the power i to get

$$i^i = (e^{i\frac{\pi}{2}})^i = e^{-\frac{\pi}{2}}.$$

5. Ernst Mach, *Space and Geometry: In the Light of Physiological, Psychological and Physical Inquiry*, trans. T. J. McCormack (New York: Dover, 2004), 104.

6. Ernst Mach, "The Economical Nature of Physical Enquiry," in *Popular Scientific Lectures*, trans. Thomas J. McCormack (Chicago: Open Court, 1895), 195–196.

7. Mach, *Space and Geometry*.

Chapter 22: Invisible Gorillas

1. Percy Bysshe Shelley, *Prometheus Unbound*, act 2, scene 2.3, lines 35–42.

2. Jerry Lettvin died at age 91 in 2011.

3. J. Y. Lettvin, H. R. Maturana, W. S. McCulloch, and W. H. Pitts, "The Mind: Biological Approaches to Its Functions," *Proceedings of the Institute of Radio Engineering*, vol. 47 (1959): 1940–1951.

4. Calvin S. Hall and Vernon J. Nordby, *The Individual and His Dreams* (New York: New American Library, 1972). Also "What People Dream About," *Scientific American*, vol. 184 (1951): 60–63.

5. Suzanne Langer, *Mind: An Essay on Human Feeling* (Baltimore: Johns Hopkins University Press, 1984), 265.

6. Suzanne K. Langer, *Philosophy in a New Key: A Study in the Symbolism of Reason, Rite, and Art*, 6th ed. (Cambridge, MA: New American Library, 1954), 58.

7. Ibid., 206–207.

8. Emily R. Grosholz, *Representation and Productive Ambiguity in Mathematics and the Sciences* (New York: Oxford University Press, 2007), 25.

9. With a bit of pride, I don't mind telling you that this interviewee was my former student Sam Northshield, who is now a professor of mathematics at SUNY Plattsburgh. Of course, a selection that includes my former student is clearly marked "suspicious" as a scientific survey.

10. Daniel Kahneman, *Thinking Fast and Slow* (New York: Farrar, Straus and Giroux), 56.

11. Keith E. Stanovich, *Who Is Rational?: Studies of Individual Differences in Reasoning* (Mahwah, NJ: Psychology Press), 126.

12. Christopher Chabris and Daniel Simons, *The Invisible Gorilla: And Other Ways Our Intuitions Deceive Us* (New York: Crown, 2010), 5–7.

13. U. Neisser and R. Becklen, "Selective Looking: Attending to Visually Specified Events," *Cognitive Psychology*, vol. 7 (1975): 480–494. Also see A. Mack and I. Rock, "Inattentional Blindness: Perception without Attention," in *Visual Attention*, ed. R. Wright (New York: Oxford University Press, 1998), 55–76.

14. Christopher Chabris and Daniel Simons, "Gorillas in Our Midst: Sustained Inattentional Blindness for Dynamic Events," *Perception*, vol. 28 (1999): 1059–1074.

15. H. Poincaré, *The Foundations of Science*, trans. G. B. Halsted (New York: Science Press, 1913), 212.

16. David Tall, ed. *Advanced Mathematical Thinking* (Holland: Kluwer, 1991), 3–21.

17. Stanislas Dehaene, "The Organization of Brain Activations in Number Comparison: Event-Related Potentials and the Additive-Factors Method," *Journal of Cognitive Neuroscience*, vol. 8, no. 1 (1996): 47–68. An abbreviated description may also be found in Stanislas Dehaene, *The Number Sense: How the Mind Creates Mathematics* (New York: Oxford University Press, 1997), 223–227.

18. T. Allison, G. McCarthy, A. Nobre, A. Pruce, and A. Belger, "Human Extrastriate Visual Cortex and the Perception of Faces, Words, Numbers and Colors," *Cerebral Cortex*, vol. 5 (1994): 544–554.

19. Dehaene, "The Organization of Brain Activations in Number Comparison," 221.

20. Masaki Maruyama, Christophe Pallier, Antoinette Jobert, Mariano Sigman, and Stanislas Dehaene, "The Cortical Representation of Simple Mathematical Expressions," *NeuroImage*, vol. 1 (2012): 1444–1460.

21. M. Cappelletti, B. Butterworth, and M. Kopelman, "Spared Numerical Abilities in Case of Semantic Dementia," *Neurophychologia*, vol. 39 (2001): 1224–1239.

22. C. Lemer, S. Dehaene, E. Spelke, and L. Cohen, "Approximate Quantities and Exact Number Words: Dissociable Systems," *Neuropsychologia*, vol. 41 (2003): 1942–1948.

23. William Faulkner, *Three Famous Short Novels: Spotted Horses, Old Man, The Bear* (New York: Vintage International, 2011), 185.

24. Jared F. Danker and John R. Anderson, "The Roles of Prefrontal and Posterior Parietal Cortex in Algebra Problem Solving: A Case of Using Cognitive Modeling to Inform Neuroimaging Data," *NeuroImage*, vol. 35 (2007): 1365–1377.

25. Ibid.

26. Anthony R. Jansen, Kim Marriott, and Greg W. Yelland, "Comprehension of Algebraic Expressions by Experienced Users of Mathematics," *Quarterly Journal*

of Experimental Psychology, vol. 53, no. 1 (2003): 3–30.

27. Anthony Jansen, Kim Marriott, and Greg Yelland, "Parsing of Algebraic Expressions by Experienced Users of Mathematics," *European Journal of Cognitive Psychology*, vol. 19, no. 2 (March 2007): 286–320.

28. F. Max Müller, *Three Introductory Lectures on the Science of Thought* (Whitefish, MT: Kessinger Publishing, 2003), 46–47.

29. Ibid., 46.

Chapter 23: Mental Pictures

1. Ludwig Wittgenstein, *Tractacus Logico-Philosphicus* (London: Routledge, 2001), 12.

2. Ibid., 19.

3. Müller, *Three Introductory Lectures on the Science of Thought*, 47.

4. Jacques Hadamard, *The Psychology of Invention in the Mathematical Field* (New York: Dover, 1954), 69.

5. Ibid., 75.

6. Ibid., 76–77.

7. This syllogism, L. Carroll, *Symbolic Logic* (New York: Dover, 1958), 118.

8. Hadamard, *The Psychology of Invention in the Mathematical Field*, 75–76.

Chapter 24: Conclusion

1. My thanks to Florin Diacu (who read this chapter in manuscript form) for pointing out to me that there is plenty of evidence to support the view that anybody who can read or write can become a very successful mathematician. I fully agree: with hard work and a stubborn will to understand and succeed, anyone who can learn to read and write can do excellent mathematics.

Appendix B: Newton's Fluxion of x^n

1. John Harris, *Lexicon Technicum: Or, An Universal English Dictionary of Arts and Sciences: Explaining Not Only the Terms of Art, but the Arts Themselves*, vol. 2 (London: D. Brown, 1723), 48.

Index

and Indian letters, 70
and Indian numerals, 61, 62, 76, 244n3
and medicine, 59
and papermaking, 98
and place-value idea, 38
and science, 59
and translations, 97
Aramaic, 103, 105
archetypes, 217
archetypical symbols, 219, 220
Archimedes, 59, 97, 123, 142
    Sand Reckoner, 20
Arethas of Patras, 86
Ariadne, 166
Aristotle, 59, 97, 210
arithmetic, 53
    and Chinese mathematics, 28, 31–32
    and Fibonacci, 66
    and Libri, 74
    and marketplace, 77
    and Stifel, 127
arithmetic statements, xi
Arnold, Bill, *Who Were the Babylonians?*, 12
Arrighi, Gino, 66
Artin, Michael, 212–213
Aryabhatta, xvi, 59, 75, 81
associative law, 188
astrology, 61, 109
astronomy, 57, 77
    Arab, 58
    and Babylonians, 10
    and Brahmagupta, 59
    and Chinese mathematics, 28
    tables for, 71, 109
    and Vedas, 36
Aubrey, John, 160
Australia, aboriginal tribes in, 8
axioms, 88, 111, 153
Aztec languages, 41–42
Aztec numerals, 23, 24

Babylonia, 12–13
Babylonians, 7, 19, 21, 24, 87, 101, 117, 131
    and Chinese mathematics, 30–31
    and counting boards, 45, 46
    and geometry, 91
    ignorance of zero, 64
    and al-Khwārizmī, 62
    number system of, 14–16

place-value system of, 65
and sexagesimal (base 60) system, 10, 38,
    62
single base system of, 23
Bach, J. S., 174
Bachet, Claude Gaspard, 83, 101, 102, 103, 105,
    106, 107
Baghdad, 11, 97
Bakhshâlî Manuscript, x, 1, 75, 76
Ball, W. W. Rouse, 59, 255n2
    *Short Account of the History of
    Mathematics*, 56
Banks, Edgar James, 13
Bayt Ul-Hikma (House of Wisdom), 1, 61, 62,
    82
Bede, *De computo vel loquela digitorum*,
    42–43
Beethoven, Ludwig van, 174
Bejaia, 65, 66
Bell, Eric Temple, *The Development of
    Mathematics*, 117
Belmondo de Padua, 120
ben Ezra, Abraham
    Sefer ha-Ekhand (*Book of the Unit*), 2, 56
    Sefer ha-Mispar (*Book of the Number*), 2,
    56
Berkeley, George, *The Analyst*, 173
Bernard of Chartres, 169
Bernoulli, James, *Ars Conjectandi*, 163
Bernoulli, Johann, 159
Bibhutibhusan Datta, 76
binomials, 128
Birch, Samuel, 101
Boethius, 39, 57, 58
    Arithmetic, 74
Bombelli, Rafael, 83, 144, 146, 191
    and Descartes, 155
    L'Algebra, 133, 134–137, 138
    and Leibniz, 165
    and polynomial notation, 157
Bornstein, Marc, 217–218
bottegas, 52–53
Brahma, 59
Brahmagupta, x, xvi, 57–59, 75, 115, 198
    Brāhmasphuṭasiddhānta (*Correctly
    Established Doctrine of Brahma*), 1,
    57–58, 60, 61, 62, 81, 109, 114
Brahmins, 109
Brahmi number system, 35, 37, 38, 39

symbols for, 87

in Thorney Abbey Computus manuscript, 48

and Viète, 145

See also exponents

prime numbers, 212, 213

priming effect, 198–199

Proclus, *A Commentary on the First Book of Euclid's Elements*, 87

proofs

and algebra, 123–125

beauty of, 220

and Chinese mathematics, 29

and cognition, 201–202, 213–214

and Descartes, 155

and Euclid, 29, 87, 88

and fundamental theorem of algebra, 139

geometric, 88, 91, 155

in geometry, xv

and Greeks, 19

and Hobbes, 164

and al-Khwārizmī, 91

in mathematics, xviii, 87

proportions, 151

psychological development, 217–218

Ptolemy, 59

Ptolemy I, 88

Punjab, 35

Pythagoras, 57, 58, 242n22

Pythagoreans, 93, 123, 148

Pythagorean theorem, ix, 13, 29, 88, 151, 155, 218, 220

Pythagorean triples, 13

Qalasādi, al-, *Al-Tabsira fi'lm al-hisab* (*Clarification of the Science of Arithmetic*), 119

Qin Shi Huang, 29

quadrivium, 53, 54

quaternions, 84, 188, 230–231

radicand, 127

radix, 122, 127, 128, 129, 130

See also square roots

rainbows, 181–182

Ramus, Petrus (Pierre de la Ramée), 116

Arithmétique, 110

Twenty Seven Books of Geometry, 142

rational numbers, 86, 94, 184, 186

real numbers, 121, 139, 158, 185, 186, 188, 189

reciprocals, 107

Recorde, Robert, xvi, 144, 163, 166

Whetstone of Witte, 83, 118, 133–134

records, 7, 9, 13

rectangles, 193–194

Reisch, Gregor, *Mararita Philosophica*, 58

res, and Fibonacci, 146

res(x), 128

rhetoric

See words

Rhind (Ahmes) papyrus, 9, 101, 116–117

right triangles, 13

Robert of Chester, 2, 60, 68

Almucabala, 110

Robson, Eleanor, 13

Roche, Estienne de la, 129

Rolle, Michel, *Traité d'Algèbre*, 156

Roman letters

written in digital order, 47

Roman numerals, 22–23, 57, 76

and Arabs, 61

and Europeans, 77

and Fibonacci, 51, 55

and Gerbertian abacus, 2

ignorance of zero, 64

and place-value system, 68–69

Romans

and abacus, 46–47, 48, 242n19

counting boards of, 46, 48, 69

and modern symbol for infinity, 163

roots

and Chuquet, 129–130, 145

and Descartes, 152–153

extraction of, 145, 152–153, 184, 194

fourth, 82

negative, 114, 122, 132

and Pacioli, 130–131

of polynomials, 129, 130, 139

and Rudolff, 82, 128

and Stevin, 139

symbol for, 122

unknown, 127

See also cube roots; imaginary roots; square roots

Rousseau, Jean-Jacques, *The Social Contract*, 174

Rudolff, Christoff, 142, 146, 157

Die Coss, 82, 128–129, 131–132